Blades in the Sky

Blades in the Sky

Windmilling through the Eyes of B. H. "Tex" Burdick

T. Lindsay Baker

TEXAS TECH UNIVERSITY PRESS

This book was set in ITC Galliard 10.5 on 12 and printed on acid-free paper that meets the guidelines for permanence and durability of the Committee on Production Guidelines for Book Longevity of the Council on Library Resources.

Designed by Jim Billingsley

Manufactured in the United States of America

Library of Congress Cataloging-in-Publication Data
Baker, T. Lindsay.
   Blades in the sky: windmilling through the eyes of B. H. "Tex" Burdick / T. Lindsay Baker.
       p.   cm.
   Includes bibliographical references.
   ISBN 0-89672-293-7 (c) —ISBN 0-89672-294-5 (p)
   1. Burdick & Burdick Company of El Paso—History. 2. Pumping machinery industry—United States—History. 3. Wind pumps—Southwestern States—History. 4. Windmills—Southwestern States—History. I. Title.
   HD9705.5.P854B873 1992
   338.7′621453′0976496—dc20   92–17903
                                              CIP

92 93 94 95 96 97 98 99 00 / 9 8 7 6 5 4 3 2 1

Texas Tech University Press
Lubbock, Texas 79409-1037 USA

Foreword

From the Great Depression to World War II, Tex Burdick and his windmillers spent their lives building and maintaining water-pumping windmills, the sky-structures that made life a reality in Texas, New Mexico, and Arizona. Water—or the lack of it—meant life or death to the rural families and ranchers who settled in the arid southwestern states the first part of the twentieth century.

This book, full of Tex Burdick's historical photos, is told by the proud windmillers themselves. They tell you why they endured the blazing sun, sandstorms, blisters, rattlesnakes, and peril to life and limb running up windmills like monkeys on a string.

They describe what they ate (and why slumgullion stew was a favorite), the clothes they wore, the tools they used, the construction techniques they employed, and why Jerry, the bird dog, was their friend.

This is a good story, simply told. As one of the windmillers said, "It was just a lot of darned hard work." Those of us who work with our hands and backs can appreciate this.

Ron Hiddleston, President
National Ground Water Association
June 1992

Preface

Volumes have been written about the "gun that won the West," but a good case could be made for the windmill and barbed wire having been equally important. It would be difficult to overestimate the role windmills played in the settlement of Texas and the West, particularly those drier regions not blessed by plentiful springs, creeks, and rivers.

Because water is essential to life, early settlement was restricted to areas that had abundant natural water or where water tables were shallow enough for hand-dug wells. The coming of the windmill, pumping water up from great depths, opened vast dry areas and made them habitable for human beings and their livestock.

For many oldtime cowboys, working on a windmill was probably their first introduction to any machinery more complicated than a coffee grinder.

Growing up on a West Texas ranch not far from the salty Pecos River, we Kelton boys ranked windmilling chores at or beneath the level of digging postholes. Windmilling was hot, sweaty, dirty work, sometimes done on an emergency basis while cattle crowded against the waterlot fence and bawled in thirst.

We lived on the McElroy Ranch, which had about eighty windmills to serve the livestock on some two hundred and twenty or so sections of marginal rangeland. The company had a fulltime windmill man, Cliff Newland, responsible for visiting every one of the mills on a regular basis, making sure they were pumping enough water to keep up with the needs of the cattle. It is amazing how much water one seven- or eight-hundred-pound lactating cow can drink in a day, especially in the heat of summer. Any breakdown could rapidly precipitate a crisis, for it would not take long for a cow herd to drink a storage tank dry. The water supply had to be restored quickly or the cattle moved.

Actual repair of the windmills was a two-man job at best, so my foreman father and the cowboys often pitched in to help. If we kids were not in school, it befell our lot to be part of the windmilling crew.

We had much rather have been riding horses.

At the end of the lowest sucker rod in the bottom of the hole, a set of molded leathers forms a seal of sorts to prevent water from draining back out of the valve instead of rising to the top. Over time, riding up and down inside the drop pipe as the wheel turns overhead, these leathers wear down and have to be replaced. This means that the sucker rod must be pulled up. It may be done by hand if the well is shallow enough. More likely the job will require block and tackle in the top of the tower and the pulling power of a horse or pickup. The rod is disconnected joint by joint until the last one is out of the hole so old leathers can be removed and new ones installed. These rods may be wet and slippery, and wooden ones can leave splinters in youthful hands not protected by gloves. The deeper the well, the greater the number of joints, the heavier the sucker rod, and the harder the job.

But replacing worn leathers is easy compared to having to pull the drop pipe, within which the sucker rod works. This requires not only a block and tackle but elevators and special gripping tools to unscrew each section of pipe from the one below it. These joints tend to rust and seal together, so that it takes muscle, sweat and grunting to release the frozen threads.

Woe unto the careless lad who allows the lower section to slip back into the hole. In such an event prayer seems of little avail, though profanity in liberal amount offers certain emotional release during the desperate fishing job that follows.

Looking back on all the Western movies I have watched, I do not remember ever seeing John Wayne or any other cowboy hero pulling sucker rod or pipe from a well. As writer John Erickson has noted about the unpleasant job of assisting birth in first-calf heifers, it is not work that fits the romantic image.

Nevertheless, keeping the mills running is one of the most vital elements in a ranch operation if other sources of water are not readily available. As boys, we recognized the necessity of the work even as we deplored it. The sight of thirsty cattle crowding around a dry water trough or bogging in the muddy remnant of a dirt tank has a way of spurring even a reluctant youngster to action.

It is for good reason that the region west of the Pecos River was the last part of Texas to be settled. It was the driest and had the least amount of available water at surface level. Without water wells, most of it would not be habitable even today.

A lone windmill, standing in the midst of an open prairie, is a Western icon of sorts, beloved of photographers and painters. A rancher or stockfarmer finds something soothing to the soul in the sight of a windmill tower, its wheel turning in the wind and clear water pumping from the end pipe onto an open tank or set of troughs.

My father seemed to have a strong feeling for the windmill. Some of it may have gone back to an accident in his childhood. When we boys were saddling up for a day's work on horseback, Dad always urged us to drink plenty of water before we left. We wondered why he insisted that we go back and take a few swallows more. He told us, finally.

When he was a youngster growing up on the Scharbuer Cattle Company Ranch north of Midland, he was often sent out with an old bachelor cowboy named Wes Reynolds, who did not willingly suffer nonsense. One hot summer day, Dad and Wes were driving a herd of cattle up the Midland-Lamesa lane. Windmills stood at intervals on the essentially level plain. Every time they neared one, Dad would lope off to get a drink of water, leaving Wes with the cattle.

At length Wes had enough. When Dad started to leave for yet another drink, Wes said, "The cattle can take care of themselves for a little while. I believe I'll go with you."

As he had been doing, Dad took a couple of quick swallows and turned away. Wes solemnly said, "I don't believe you've had enough. You'd better drink a little more."

Dad was content with another swallow, but Wes would not let him quit. He made him keep drinking until he was sick at his stomach. Then Wes declared in a gravel voice that meant business, "Next time, damn you, you'll *water out* before you leave the house!"

Dad always saw to it that we *water out* at the house or windmill before we proceeded to the job, whatever it was.

No water ever tasted better than when it came up clear and cool from deep in the ground, its flow pulsing to the steady rhythm of the wind-driven pump.

As a small boy I was stirred to awe by the tall tower up in whose ladder I was forbidden to set a foot, and the huge, unreachable wheel high above, which always knew from what direction the wind was blowing. A story of a young cowboy killed by a falling sucker rod filled me with a sense of dread and for a long time kept me from venturing beneath the outer braces.

Many of the McElroy windmills were still the old wooden Eclipses, which required frequent oiling to keep the bearings from freezing up. Every wooden tower carried a congealed coating of spilled oil, which not only made the ladder slick and treacherous to climb but left its indelible mark on clothes. These Eclipses were gradually being replaced by the newer and more efficient self-lubricating steel types made by Aermotor and other manufacturers. By the time I was in my early teens, the last of the huge old cypress fans lay weathering in the sand, though many of the wooden towers remained, topped by dependable steel wheels which turned easier and pumped as much or more water than the old ones had.

Today, where electrical power is available, many a windmill tower lies on the ground or stands in skeletal ruin, replaced by submersible pumps. But much of the range country is still far from power lines, and the windmill remains essential for domestic livestock as well as steadily increasing numbers of wildlife. Across much of Texas, because ranchers and farmers have sought to protect and propagate deer, antelope, wild turkey, quail and other game, many wildlife species are as numerous as when the Indians still held title. For these, as well as for cattle, sheep and goats, the windmill remains a life-giving force in an environment that otherwise might be impossibly hostile.

Windmill men such as Tex Burdick and others described in T. Lindsay Baker's narrative deserve much credit for making life possible in semi-desert rural areas of Texas, New Mexico, and other parts of the West.

Elmer Kelton

Introduction

Blades in the Sky describes the work of a team of Texas-based windmill erectors in the first half of this century. The text has been woven around a remarkable private collection of contemporary photographs that depict typical scenes from windmill construction sites in the 1920s and 1930s. In the broad sweep of United States regional economic development and the development of ground water resources, the significance of *Blades in the Sky* goes far beyond Baker's immediate narrative and the Burdick photographs.

My first contact with Dr. Baker goes back to the mid eighties when I corresponded with him concerning an interest he had in the development of windmills in Southern Africa. As a hydrogeologist my interests were principally geological, and my only contribution to his request at that time was to point out that mills milled and pumps pumped and that therefore his real interest was in wind pumps not windmills! I haven't managed to change his or anybody else's usage of the term *windmills* but our correspondence eventually led to me making trips to Texas where I have enjoyed Dr. Baker's hospitality first in Canyon and more recently in Waco.

There are two strong common interests shared between Lindsay Baker and myself. Firstly, we are both educators, and recognize the fascination of the past as a means to stimulate and focus inquiring minds on the present. The second common interest between the "hydrological earth scientist" and the "historian" is related to a fascination with the role of technology in developing natural resources. In the development of the west in the last third of the 19th century, three separate but related technologies worked together: the development of water-well drilling technology to reach beyond the very limited depths of hand-dug wells; the development of windmills to use the force of the wind to pump water from underground; and the invention of barbed wire and the means to mass-produce it. These were three interrelated ingredients of the development of the southwestern United States, each one critical in transforming cattle-based agriculture. Without fencing, there could be no selective breeding; without drilling, there could be no economic access to range lands away from permanent water courses; without wind-powered pumps, there could be no access to vital ground water resources. The early development of many ranch homesteads, small towns, and the railroads in the United States were totally dependent on windmills to raise water from aquifers.

Blades in the Sky principally concerns Texas and a small but critical service industry that evolved to apply technology to provide the vital link between ground water resources and water demand. Windmillers were a select band of specialists who installed and erected windmills and water pumps.

In a much wider context, much of the late 19th- and early 20th-century development of windmills in the west and south of the United States was in part dependent on initial inventions and technical developments of the windmill by inspired small-town entrepreneurs and engineers from New England. A short digression to earlier times, and to different places, will give a little background and context to the craft and the art of the Texas windmillers.

The first recorded windmills in the United States were in Jamestown, Virginia in the 1620s and were used for grinding flour. The development of wind power for raising water is a more recent application and has had a profound effect in rural development in the United States, as well as in countries throughout the world. Engineers had long ago solved the problem of transforming the circular motion of a wind-driven sail to the reciprocal motion needed to drive pumps, but what was missing until the mid 19th century was a means to control automatically the speed of the turning blades so that the windmill and the pump would not self-destruct in high winds. On August 29, 1854, the United States Patent Office approved the "self-governing" windmill invented by Daniel Halladay of Vermont. Production soon moved to the midwest, where there was a greater demand for wind-powered pumps, and by 1855 the Illinois Railroad Company of Chicago was a early Halladay windmill customer. In Halladay's design, an increasing strength of wind changed the pitch of the blades as they faced the wind. The change in pitch slowed the wheel revolutions. A different concept, patented by Leonard Wheeler in 1867, used a rigid mill wheel with a vane behind. In high winds, the vane forced the revolving wheel away from the main wind direction resulting in a slower rate of turn for the wheel. The original Eclipse windmills used this design.

Many companies were involved in United States windmill manufacture in the 19th century. There was tough competition and an almost endless advance in improvements and modifications. Patent lawyers were kept busy as manufacturers improved bearings, gears, self-oiling mechanisms, blade design, etc. The evolution of windmill designs, the background to inventions and the registering of patents, and the fierce commercial competition and company takeovers, make an incredibly interesting aspect of American history. The definitive book on the subject is *A Field Guide to American Windmills* written by Dr. Baker in 1985. Most Americans who travel by road in rural areas will be familiar with two or three of the famous name windmills such as Dempster, Eclipse, Monarch, and Fairbanks. There are, however, hundreds of different manufacturers and a plethora of models that have been produced in the last hundred years. The *Field Guide* is a mine of information that has been meticulously researched, much of it from hitherto-ignored contemporary trade publications. When seen in the context of the development of American windmill technology and manufacturing, the importance of *Blades in the Sky*, as an original cameo of American history, becomes especially significant.

The increased application of wind power for water raising throughout the west in the last century necessitated the evolution of service-industry entrepreneurs to build the windmills and service the pumps. Once a settlement or range was dependent on wind-raised ground water, it was essential to have a fast and responsive repair service. Lack of water could mean death of cattle and economic ruin for the landowner. There do not appear to be published contemporary accounts of the 19th-century windmiller artisans and businessmen, but probably the 19th-century activities of windmillers were not too different from the scenes captured in the Burdick photographs.

By the 1860s and 70s, there was a great demand for windmills, a demand that was paralleled in the barbed-wire industry by demand for economical fencing. Proliferation of windmill production and barbed-wire manufacture (there are 900 or so different designs of barbed wire) is an indication of the surge of demand and output in the farming community as the short-lived "wild west & cowboy" era ended and development began toward scientific farming and today's agribusiness.

Windmills were produced in many different sizes and configurations depending on the windiness of an area, the depth of pumping necessary, and the amount of water needed. Many windmill manufacturing companies had a "mix & match" inventory, which required the windmiller to become an advisor and consultant to farmers about what design was best suited to a particular need. The owner of a windmill installation company usually provided far more of a service than merely to erect a tower and install the sails and pump.

Until the use of rotary drill in the 20th century, virtually all deep water wells were drilled by cable tool rigs, many of which were home made, of timber frame construction, and which used horse power as the means of raising the drill bit. Some shallow wells were constructed by hand digging, and in some valley situations, wells could be driven up to forty feet into sandy sediments. (The story of early well drilling in the United States is not well documented and should perhaps be the subject of research for Dr. Baker's next book!)

In any process involving many stages, as in a machine with many cogs, each stage, or cog, is a vital part of the total process. In the development of ground-water sources by the use of windmills, the process of assembling the sails and gears, designing and erecting the tower, and installing the pump are too often taken for granted, or perhaps not really ever considered. The windmiller has been, however, a vital and hitherto unsung "cog" in the process of exploiting ground-water resources for over a hundred years. There are still firms today that offer the service of erecting and installing windmills. The majority of today's "windmillers" are likely to belong to a family-owned well drilling or pump-installing business that offers a wide range of additional water supply services to customers.

The trials and tribulations, success and enterprise of the Burdick company, as documented in *Blades in the Sky*, serve to exemplify the family business ownership that is still characteristic of much of the current United States water-well contracting industry, even though windmill power for pumping may have been replaced by electrical submersible pumps and horse-powered drilling rigs by $500,000 monster drilling machines. The professional services of today's water-well contractor are highly technical, including, for example, advanced designs in high production wells and precisely installed sampling wells used in scientific studies of contaminant migration. The vital personal service in providing an application of technology to satisfying water needs, however, remains the same as it was in the heyday of the windmillers. In the 1990s, windmills may be a small part of the United States ground-water industry, but there is little doubt that for some areas and for some water needs, wind pumping will continue to be the most effective, economical, and environmentally friendly way of linking economic need with the vital underground source of the nation's water. With over 50% of the United States population dependent on ground water for drinking purposes and more than 13 million private drinking water wells, the United States water-well industry of today is as much a vital cog in national economic development as were Burdick's men in the development of rural Texas in the 1920s.

All travellers are familiar with the silhouette profile of a slowly turning windmill in a rural landscape at sunset. How many have ever considered the complexities and difficulties of installing the windmill? How many have understood the windmill as a symbol of the nation's underground wealth? *Blades in the Sky* provides a fascinating insight into one aspect of the process of the development of water resources.

No reader of *Blades in the Sky* will ever again see a windmill (or wind pump!) in the landscape without giving thought to the sometimes harsh and semi-nomadic life of the windmillers. The survival of Burdick's photographs and their publication along with the accurate historical context of Baker's text provide a wonderful insight to times past. With only a little imagination, the reader of *Blades in the Sky* can be moved back in time, and extended in geography, far beyond Burdick territory. Read on . . . .

Andrew Stone
Program Director
American Ground Water Trust

To Billie Wolfe
who introduced me to research on windmill history.

Blades in the Sky

The Story

"I worked for Tex Burdick many a day for two dollars a day and all I could eat,"[1] remembers former windmiller Harry J. Clifford. He shares the memory with many other windmill men, for in the 1930s and 1940s Burdick's firm in El Paso, Texas, became one of the largest distributors of windmills in the desert Southwest, employing multiple crews and selling as many as eight railroad carloads of windmills annually.

Windmills became a virtual gold mine for B. H. "Tex" Burdick half a century ago because they had become necessities for ranchers in the arid and semiarid Southwest. The technology of wind-powered pumps was well established by the 1880s, and within a decade windmills dotted the regional landscape, towering above the scrub brush and sparse grass. Before the introduction of drilled wells and windmills to pump them, ranchers could run cattle on only those ranges that had water from dependable streams or springs. Most of the upland grazing areas were beyond their reach. Individuals who could command the few water holes could effectively control the hinterland.

The introduction of technology for drilling wells and then pumping water with the free power of the wind wrought great changes in ranching. For the first time, stockmen could used barbed-wire fencing to divide their ranges into individual pastures, each watered by a windmill. Frequently this permitted their first real planned breeding of stock. With more cattle on the upland ranges, the stockmen depended increasingly on windmills for their water supply. Cattle could go only three days without water, so if a well ceased pumping the rancher had to take immediate steps to secure water by returning the well to service, moving stock, or hauling water. With such a dependence on machines to provide water in a potentially hostile environment, Tex Burdick's windmills and emergency maintenance became essential components of his customers' livestock-raising operations. For many stockmen, Tex Burdick even rivaled in importance the role of the local veterinarian.

Tex Burdick held his rancher customers in the highest esteem. A stockman might come into Burdick's office and say "I want a windmill, but I can't pay for it until I sell my cattle in the fall."

[1] Background data for this book and sources of all the windmillers' quotations cited herein are in the following tape-recorded interviews and their typescripts at the Research Center of the Panhandle-Plains Historical Museum, Canyon, Texas: B. H. "Tex" Burdick, Sr., to Larry D. Sall, El Paso, Texas, December 12, 1975; Burdick to T. Lindsay Baker, El Paso, March 6, 1981; Burdick to Baker, Rio Vista vicinity, Texas, March 14, 1991; and Harry J. Clifford to Baker, Albuquerque, New Mexico, June 16, 1991. For further information on the history of wind-powered pumping and its applications in the desert Southwest, see the readings recommended at the conclusion of this book.

This might take place in April or May, but, according to Burdick, "you knew when he sold the cattle, you were going to get your money. . . . if he shook hands on it, this was as good as him . . . signing an agreement in front of a notary."

When Burdick entered the windmill business in the 1920s, he had for a crew only himself and a couple of helpers, but in time he added employees until on the eve of World War II he operated three three-man crews in the field erecting and repairing mills. The work procedures for these crews remained pretty much the same for the two decades that Burdick dealt in windmills.

The men on the crews saw themselves proudly as "windmill men," even telling city directory fieldworkers that they were "erectors." They developed their own small vocational subculture that extended northward from Texas to the Canadian prairies. All of them were proud of their work and their ability to undertake jobs high above the ground that daunted ranchers, ranch managers, and cow hands, all of whom otherwise viewed themselves as men's men. Most of these people on the ground could by no means muster the courage to climb around on windmill towers "like a monkey on a string" the way that Burdick's crews did.

For Tex Burdick's men, the days were long and the calendar meant nothing. They considered themselves fortunate to have steady jobs during the days of the Great Depression. "We worked 'em from sunup till sundown," Burdick states, "In the summertime that was pretty long." Whenever there was work to be done, the crews were either in the field or on the road between El Paso and the installation sites. Work went seven days a week whenever it was to be had, although in slack times the weekends were usually free. A crew could erect and install a thirty-foot steel tower with a smaller eight- or ten-foot-diameter windmill over a shallow well in the El Paso valley in a full day, but the larger mills with heavier towers, which usually were in more distant locations, generally required three days. Jobs that involved one of the truly large twenty- or twenty-two-foot-diameter mills or perhaps a circular

4

*Steel Eclipse
Type WG
windmill erected
by Tex Burdick
in Mexico, circa
1930*

steel water storage tank frequently required a week in the field. The Burdick territory extended from trans-Pecos Texas through southern and central New Mexico into southern and eastern Arizona.

Crews generally consisted of a top man and two helpers. The top man supervised the job and was the person responsible for most of the work above the ground on the tower. Because of his greater responsibilities, the top man received five dollars a day, contrasted with two dollars a day for the men on the ground. All employees received their food while on the job as a benefit of employment.

Because both Spanish- and English-speaking employees comprised the crews, bilingual communication was essential for Burdick's men. Most of them had spoken both languages from childhood. Harry Clifford grew up with Spanish-speaking children on an El Paso Valley dairy and spoke as much Spanish as English by the time he started to school. Others, Tex Burdick, for example, had learned their second language through formal education. Whatever the level of bilingual fluency an employee might have had at the start, he soon expanded it through daily interaction with co-workers. Speaking only his "high-school Spanish," Tex Burdick, early in his windmilling career, traveled into Mexico to install a mill at the Coralitos Ranch across the border from Columbus, New Mexico. Before the installation was completed a freak snowstorm forced him, his three Mexican helpers, and a Mexican cook to halt their work. "For three days we had to sit in a tent," he relates. At the outset Burdick had spoken only halting Spanish, but after several days snowed in with the other four men he could do much better. "By the end of the three days, I could even sing some Mexican songs," he says.

One of Burdick's crews might leave El Paso for the field at any time of day. All depended on when they got their vehicle loaded. For local installations in the El Paso Valley, a passenger car towing a trailer might carry everything that was needed, but for larger mills in more distant locations, the crews used trucks sometimes pulling trailers. The men would load their tools and gear onto the vehicles outside the Burdick building at 190 North Cotton Avenue in El Paso, manhandling the components of the mills and tanks, using an overhead crane in later years for the heaviest pieces. Then they would head out, all three men in the cab of the truck or in the passenger car. Sometimes during the summer heat the crew members would wait until dusk to leave so that they could drive in the comparative coolness of the desert night.

If the destination for the crew were in western or central New Mexico or in Arizona, their first stop usually would be in Las Cruces, New Mexico, about forty-five miles up the Rio Grande Valley from El Paso. "There was one cafe there," remembers Harry Clifford. "We always stopped there and had a big steak dinner or whatever the boys wanted to eat. Then we'd go on." If the crew needed fresh groceries they would buy them in Las Cruces or in the town nearest to the installation site. "We'd . . . stop [at] some little store along in there and buy it and go on to the job site," Clifford says.

On arrival at a ranch, the top man would confer with the owner or ranch manager about the location of the well for the installation or repair. "It might be twenty miles out there to where the windmill was going," Harry Clifford notes. When the crew reached the site, they might see only a piece of six- or eight-inch pipe sticking out of the ground a couple of feet with a metal cap or a rock on top of it as the driller had left it. If they arrived in the evening, the men would pitch their camp, build a fire, cook supper, put on a pot of beans for the next day, and lay out their bedrolls for the night.

6

Tex Burdick provided his crew with tarps and light cotton army surplus mattresses. The men brought their own blankets and other bedding. In that environment, rain was so rare that Burdick never provided tents. Harry Clifford recalls that beds got soggy from time to time, but that "most of the time we got along real well."

The men slept wherever they wanted so long as it was on the ground. "We never . . . slept together," notes Clifford. "Everybody kind of slept where they felt like it would be the best place." Most of the crew followed the same basic procedure in laying out their pallets. "When you bedded down at night, you saw which way the wind was from," Burdick explains. Then, "you put your feet into the wind and pulled the tarp up over your head and went to sleep." If the wind changed during the night and began whipping up beneath the covers, the sleeper simply got up, grabbed one corner of the bedroll and tarp, turned the foot of the bedding to the wind, and retired again.

About daylight the top man roused the remainder of his crew and started preparing breakfast. Sid Bowlin, one of Burdick's most memorable top men, gathered his firewood and cow chips each evening and put them under a gunny sack so that dew or moisture would not collect on them. Then the first thing in the morning, he would lean out of his covers to light the campfire before dressing. After breakfast the crewmembers rolled up their bedding and tarps "to keep the bugs and stuff out," remembers windmiller Harry Clifford.

Sid Bowlin set the pattern for handling tools in the field. "When he arrived at a mill site," Burdick relates, "he would lay out a gunny sack, take his tools out of an old ore bag, and lay them down very carefully on this . . . sack because you can lose tools at a windmill site quicker than anything

else." The helpers all knew to keep the tools in this one location unless they were being used.

In the installation of a windmill, the first step was to dig the holes for the anchor posts. For small- and medium-sized mills, these holes were five feet deep, but for larger mills they were six feet deep, all big enough for a man to get inside and dig. Everyone on the crew excavated anchor holes. There were no exceptions to the rule, including B. H. Burdick as the boss if he happened to be on the job. "You dug your own hole," he remembers, "Nobody else helped anybody else."

Once the anchor holes were prepared, the crew members set into them the anchor posts to which tower legs would be attached. In especially alkaline soils, the men encased the metal anchor posts in concrete, but otherwise they filled in the holes with rocks and soil, tamping them down securely. It was extremely important that the anchor posts be precisely square and level, for if they were not exactly where they were needed the tower legs would not come together evenly at the top.

The tower might be assembled in one piece lying on the ground or built up from the ground one piece at a time. The decision on the erection method was based on a number of factors, among them the size of the tower, the amount of space available at the site, and the availability of other vertical elements that might be used in hoisting the assembled tower with steel cables. Because larger towers were more difficult to raise from a vertical position, derricks fifty feet or taller were usually built up from the ground.

If a tower were raised in one piece, the main casting and vane assembly of the mill usually were raised with it, leaving only the wheel to be installed. Sometimes even the wheel was raised with the rest of the mill. On the tower built up from the ground, the crews manhandled the

Windmillers (from left) Tony Venagas, Harry Clifford, and Carl Boyd having breakfast in camp at the Warner Ranch near Rodeo, New Mexico, December 1941

components of smaller mills into the air with ropes. For mills ten or more feet in diameter, the men used gin poles with blocks and steel cable tackle to lift the parts to the tops of the towers for installation. Once the mills were assembled, the crews generally installed pumping cylinders, pipe, and sucker rod in the wells and connected the mills so that the systems would pump water.

Despite popular misconceptions, the windmill itself does not actually pump water. The windmill atop a tower converts the force of the wind into rotary motion with its wheel and then translates that rotary motion into a useable up-and-down motion. The sucker rod transmits this up-and-down motion underground to a simple piston-displacement pump called the pumping cylinder. From its position underground at the water table, this pump pushes the water up the pipe to the surface for use.

Tex Burdick's crews did more than install windmills. Frequently they were called to the field to render emergency repairs on customers' water systems. They installed and repaired not only windmills but also pumping equipment operated by electricity and internal-combustion engines. Occasionally customers asked the crews to construct large circular steel water-storage tanks adjacent to their wells to hold reserve supplies for times of calm or when the wells might not be in service. Made from six- by twenty-foot pieces of second-hand sheet steel material, which earlier had formed oil field storage tanks, these twenty- to fifty-foot-diameter reservoirs required considerable effort to build, but the crews took the work in stride. "It was bolted together," Harry Clifford remembers, "about a million bolts to put it together, one man inside and one out, tightening bolts."

Although compensation of two dollars a day for semi-skilled labor and five dollars for supervisory employees seems paltry today, during the depths of the Great Depression B. H. Burdick's men were pleased with their wages. "Back in those days," Harry Clifford states, "it was . . . one of the better jobs around, 'cause . . . there wasn't much work going on . . . and we always knew we were going to eat good, no matter where we were."

During the hard times of the 1930s, food indeed was a large inducement to join one of the Burdick windmilling crews, for wherever the men were working, either in town or in the country, Tex Burdick provided the meals. "If we were in town working for him," Clifford notes, "our meals were paid for." When on the job in El Paso, many of the men dined around the corner from the Burdick building at the Rock Hut Cafe at 1612 Texas Avenue. If they were in the field, the employees prepared their own meals over a campfire but could choose at the grocery store whatever foods they preferred. "We didn't have to worry about . . . him hollering about it, because he paid for whatever we bought," Clifford remembers. The provision of meals, in fact, prompted Clifford to join Burdick's crew, for in 1936 he was working in El Paso for the Western States Grocery Company at $17.50 a week, but there he had to buy his own food. "I got crosswise with them and quit my job," he relates, "and Tex gave me a job. . . . I worked for Tex Burdick for fourteen dollars a week, and I had all I could eat." When top man Carl Boyd joined Burdick's crews, he had been working in a warehouse for the Bigelow Supply Company, wholesale confectioners at 611 East San Antonio in El Paso. "He was getting by on nothing," Harry Clifford remembers. "He was always munching on a candy bar. He always had a carton of candy open somewhere, and that was his dinner and his lunch." For Carl Boyd the idea of having three square meals a day must have been a considerable enticement to change jobs.

Food for Tex Burdick's men in the field must have been good, for to this day former members of the crews talk about how much they enjoyed it. Top man Sid Bowlin, a former ranch hand, was the star cook, and all the laborers vied to work on his crew so that they could enjoy his meals. Harry Clifford remembers especially Bowlin's biscuits baked in a Dutch oven: "He would take and heat it up. Make up these biscuits and place 'em in there, and put the lid on and the coals on the lid. He knew just when to take 'em out." Tex Burdick agrees: "He could make bread in a Dutch oven that was just out of this world."

9

Jerry in camp with Tony Venagas on the Warner Ranch near Rodeo, New Mexico, December 1941

There were two initial steps in the preparation of food at one of the Burdick windmiller's camps. One member of the crew started a campfire fueled by cow chips, scrub brush, or pieces of windmill shipping crates and put on a big two-gallon pot of coffee, while another dug a trench about a shovel wide, six inches to a foot deep, and a foot and a half to two feet long. With slats and a wet gunny sack spread across the top, the pit served to keep cool such perishables as lard and bacon. The men never had the luxury of butter in camp, a lack that they all felt.

Breakfast for a Burdick windmilling crew in the field typically consisted of either fried or

10

scrambled eggs and toasted bread or biscuits, sometimes supplemented by bacon or potatoes. As soon as a camp was established, one of the men would pour water into a pot with dried pinto beans and start them simmering on the fire, so by breakfast or lunchtime on the first full day in the field the beans seasoned with chili would be ready to eat. "Beans . . . was the most important item that we had on our menu," Tex Burdick remarks. Occasionally the breakfast routine might be altered by the substitution of pancakes grilled in the bottom of a Dutch oven and topped with Karo brand corn syrup.

The noontime menu varied with the season of the year and the length of the job. During hot weather in the desert, the men often ate as simply as sandwiches and canned pork and beans or fruit, but in the winter the lunch might be a big pot of stew. Always beans were available as part of the meals.

The evening meal was generally the most substantial of the day and it might offer more variety. Broiled steak and either fried or baked potatoes made a favorite dinner for Harry Clifford, who remembers the men cooking their own steaks on green mesquite branches turned slowly over the coals of the campfire. The men themselves occasionally shot rabbits for fresh meat in camp. For a meal of jackrabbit or cottontail, they would usually cut the meat into small pieces and fry them in lard or bacon fat. Ranch owners and managers from time to time gave the windmill men the meat from larger wild game animals or from their own livestock. "Occasionally we would get sheep," comments Tex Burdick, "and I didn't like sheep one little bit." He continues, "I can still smell some of these sheep that I put up mills for, especially Mr. Lovelace up there at Corona, New Mexico. Man, that was a bad-smelling place!"

One of the treats that Burdick crews could look forward to was Sid Bowlin's slumgullion stew.

The former ranch hand had perfected the concoction during his cowboy days. The main elements of the slumgullion were some type of meat with potatoes and onions and chili as seasoning, but Bowlin added to those main ingredients anything else that might strike his fancy. "Meat, eggs, potatoes, corn all in that Dutch oven stirred together," remembers Harry Clifford. Tex Burdick observes, "He'd take maybe two or three jackrabbits . . . or if it happened to be quail season, I'd always have some quail. He'd put it in a Dutch oven and cook it all morning or all afternoon." When it came time to eat, "you'd ladle it out with a tin cup because the meat would fall off the quail or the jackrabbits or whatever you were using as meat . . . That was the best slumgullion stew."

The windmilling crews also carried with them a wide range of canned and dried food. Pork and beans was a favorite among the men, for they enjoyed beans and could eat these from cans with no cooking. Tex Burdick bought canned items such as pork and beans, corn, peaches, and condensed milk by the case, and usually they were on hand at the building in El Paso. Dried fruit, including apricots and prunes, also was a favorite with the men, especially the Hispanic employees. Tex Burdick remembers that one of the Mexican employees "took a sack of this and sat in the front seat there and began eating these dried fruits . . . he just sat there and kept on eating and eating . . . And when we got to the job, he got a drink of water." The consequence was that the fruit began swelling in his stomach and produced painful results. "I thought that we were going to have to take him to the doctor and get him pumped out," Burdick says.

The foodways of the predominantly Mexican American members of Burdick's labor force had their influence on the diet of all the workers. Chili was used as a seasoning in many of the dishes, particularly in beans and stews. Of the beans, Tex

View from the top of a windmill, circa 1935

Burdick remarks, "They always had enough chili in them to where they were very palatable." On one occasion, he traveled to repair an eighteen-foot mill on a ranch near Fort Davis, Texas, and he ate with the Hispanic sheepherders that the rancher had assigned to help him at the well. "They had a *caldido,* a soup over there," he remembers, "that was the hottest I ever ate. And I mean that's what we had for dinner and we had it for supper."

The principal drink for the windmill men was coffee, from the first lighting of the campfire until the breaking of the camp. Harry Clifford remembers that the men drank coffee "all day long" and

12

that they constantly kept the pot simmering. By the end of a day, he notes, the beverage was so strong that "it could stand up and fight back." Alcoholic drinks, by contrast, did not find a place in Burdick camps. While at work "we just didn't drink," Clifford reports, "You couldn't drink and walk out on the tails of those windmills."

Burdick crews used ranching-style chuck boxes for transporting their utensils and food into the field. Hauled in the back of a truck or on a trailer, the wooden box with hinged front held food-related materials in one convenient location, a rubber gasket on the opening keeping it comparatively clean. Any crew that left for more than one day carried its own chuck box. "It was just part of the equipment we had to have," Burdick states.

A constant companion for the crews whenever Tex Burdick accompanied them was the boss's bird dog, Jerry. "That dog was with him wherever he went," remarks Harry Clifford, and the statement was no exaggeration. "He went every place with me," Burdick says. "There was no question about it." More than just a friendly lick and nuzzle to the men, Jerry performed a worthwhile service as a guard dog for the camps at night. "You rolled out your bed on the ground," Burdick relates, "and old Jerry slept under the tarp but on top of the bed," ready to ward off any intruders on either two or four feet.

One evening near Rodeo, New Mexico, Burdick drove along on the railroad tracks about a quarter mile from town to put down his bedroll and sleep till dawn. "And along in the night, why, Jerry growled," Burdick says. He grabbed the dog by the nose to keep him quiet while he grappled in the bedding for his pistol and surveyed the darkened horizon. "Pretty soon I saw the silhouette of a man out there." He called out to the stranger, "Is there something that you want?" The answer was "No," and then the figure began running away. "I had to hold Jerry 'cause he would have caught him," he says.

Another way that Jerry earned his keep was in alerting the windmill men to the presence of intruders like rattlesnakes or skunks. Rattlesnakes particularly concerned the members of the work crews, who never seemed to become accustomed to finding them. On one occasion an erection crew undertook a job near Coolidge, Arizona, not far from the old site of a house of which only the basement remained. One of the men in curiosity went down the steps into the basement area, found a rattlesnake, and retreated. He returned and killed the rattler, but found more. The basement was full of tiny rattlesnakes. "He said that he had never killed so many little rattlesnakes," Burdick recalls. Finally the men decided to leave the basement full of snakes alone. "They were afraid to go down in there," Burdick says. "They even . . . put a piece of sheet metal . . . at the top of the steps there, because they were afraid that the rattlesnakes might come out at night and get in bed with them."

From time to time the vipers did indeed become bedfellows with the windmillers. "You can jump from a laying position," Burdick declares, "and if you don't believe you can do it, just find one in bed with you." The nearest that Burdick came to being bitten by a rattler, however, was on a job near Fort Stockton, Texas. The drinking water for the crew came from a water line with a faucet in a valve box just below the surface of the ground. "It was after dark and I went out there and started to put my hand down in there and unscrew this valve," he relates, but "something told me not to." He walked back over to camp, took a flashlight, and shined it down into the box. "Here's a rattlesnake coiled right around the valve," where it had sought the coolness. "It sure did shake me up, I'll tell you."

The most serious dangers to the windmill men came not from rattlesnakes or intruders in the night, but rather from injury on the job. Tex Burdick's windmilling crews remained remarkably accident free, even though they did work at considerable height and frequently moved very heavy objects. "We always looked out for each other," Burdick remarks, and former employee Harry Clifford comments, "We tried to stay pretty careful." Even so, many of the crew members undoubtedly endangered themselves as they took shortcuts to save time and effort. "They were up there doing things the easy way—getting right out there and doing it," Burdick says.

The wind, for example, rarely slowed down the erection of windmills, even though it might present real dangers. When assembling a wheel atop a tower in the wind, one man in the air held a rope pulling up a wheel section, while another on the ground held a second one pulling downward so that the sheet metal blades would not catch in the wind, fly out, and bang back against the tower. Describing that operation, Burdick says that once the man on top manages to grab the wheel section, "you wrap a leg around the corner and hang onto it till the other man gets on the other end of it and you try to get it up into the windmill arms." "Some damn things are foolish I did, . . . " Clifford comments, "but we got away with it."

Byrl H. "Tex" Burdick, Sr., born in San Angelo, Texas, on September 25, 1900, grew up in the water supply business. His father Lee Burdick had come to Texas first in 1898 as an employee of the Challenge Wind Mill and Feed Mill Company, maker of windmills and water supply equipment in Batavia, Illinois. As a young man the elder Burdick had participated in the construction of a fifty-thousand-gallon water tank for a customer of the firm in Dallas. Lee Burdick decided

that he liked Texas, and after returning to his home in Minnesota to marry his sweetheart, he came back to Texas with his bride to locate in San Angelo. There he worked for the Hagelstein Hardware Company installing pumps, windmills, and related equipment, becoming noted locally for his early use of a motorcycle. Tex Burdick relates, "He told me many times, when he crossed the Pecos River, he would put this little light motorcycle on his shoulder and wade the river and carry it to the other side and then he would go back after his tools."

In 1902 Lee Burdick with his wife and his toddler son relocated in El Paso, and it was there that Tex grew to his adulthood. During his senior year at El Paso High School, he lost his mother. The household broke up after her death, and the new graduate of 1918 decided to follow the wheat harvest north for the summer. Ending up at Faribault, Minnesota, he met an exsoldier named Ewart Austin Earl Smith, who had served in the U. S. Army at Columbus, New Mexico, during the recent border troubles involving Pancho Villa. "I was homesick and lonely and . . . he was from home," Burdick remembers. Smith planned to attend the University of Minnesota that fall, so Burdick decided to do the same and room with him. The young Texan could be distinguished from his midwestern classmates every time he opened his mouth and soon his fellow students were calling him Tex. The nickname stuck for the rest of his life.

After a winter in Minnesota during which he spent more hours earning a living than sitting in classes, B. H. Burdick decided in spring 1920 to return home. At least he knew that the weather would be warmer in the desert Southwest. Finding no work in El Paso, the twenty-year-old made his way to Hanover, New Mexico, where he worked in the zinc mines of the Empire Zinc Company and the Colorado Fuel and Iron

14

Company. It was at this time that he purchased a No. 2-C Kodak Jr. folding camera and with no formal training began making photographs. "I remember that I wanted some pictures up in that part of the country," he says, "because that was beautiful country."

When zinc prices slumped after World War I, the mines at Hanover closed, and young B. H. Burdick found himself back in El Paso, where his father had remarried and was working for a hardware company. "In desperation I worked for the city at no pay to get experience in drilling wells," he recalls. Just to learn about a certain type of well drilling, he went with a friend each morning to a site on Second Street where a municipal crew was sinking a well with a rotary

drilling rig. After a couple of months the two young men built their own jetting-type well rig on the stripped down running gear from an old Mack truck. Before long they had become well drillers in the area of comparatively shallow groundwater in the El Paso Valley.

About this time Lee Burdick, Tex's father, left the hardware company to manage a cotton gin just upstream on the Rio Grande at La Mesa, New Mexico. He was busy at the gin from midsummer to midwinter, but the remaining six months of the year he used his long-developed contacts in the ranching community to find well and windmill work for his son's well drilling operation. "He . . . would go out to these ranches and . . . sell . . . barbed wire or equipment. He would make

Burdick & Burdick windmillers posed for a photograph during the fitting of the mast pipe into the main casting of a sixteen-foot-diameter Challenge 27 windmill on the Slaughter-Veal ranch northeast of Fabens, Texas, circa 1935

15

arrangements for me to pull the wells or . . . to work on a windmill," Burdick remembers. For the next few years father and son collaborated, Lee drumming up the jobs and Tex doing the work.

In 1927, as the business started to grow, Tex rented for twenty-five dollars a month a wood-framed corrugated sheet-iron barn at 190 North Cotton Avenue from the Momsen-Dunnegan-Ryan Company, one of his major wholesale suppliers. The business increased gradually until he could employ two Mexican helpers who remained on call whenever they might be needed to go into the field. "When we'd leave there, we'd put a chain around those two-by-four doors and a lock on the outside, and we might not see that place for two, three days or even a week," Burdick remembers.

The same year that Tex Burdick rented his tin barn on North Cotton Avenue, he had new invoices printed up and needed a name to put at the top. When he had begun his small-scale water-well drilling a few years before, he had used the name Burdick & Burdick, so he decided to use it again, as he says, "because I had a son by that time." Customers already had begun telephoning Lee Burdick's residence wanting Tex to do jobs for them, so Tex asked his stepmother to go down to the barn for a little each day "and possibly answer that phone." Before long they hired a bona fide bookkeeper and stenographer to sit in the building. "There wasn't much stenographing because nobody was there, but she did answer the phone," Burdick says. "She read and knitted and didn't have much to do." Business grew, however, bit by bit as Tex Burdick sold and installed windmills and other water supply equipment that he had purchased wholesale from Momsen-Dunnegan-Ryan and the Zork Hardware Company. All this took place despite the stock

market crash in October 1929 and the beginning of a nationwide economic depression.

Finally in 1931 a representative named Redman from the Challenge Company of Batavia, Illinois, called on Tex and Lee Burdick in El Paso. This firm had succeeded the one that Lee had worked for back in the 1890s. Redman told Tex, "I understand you are putting up most of the mills in this part of the country," to which the young man responded in the affirmative. Redman asked, "Could I interest you in a small carload of Challenge windmills?" Flabbergasted by the unexpected offer, Burdick replied, "Well, I would like it very much, but I can't pay you until I put 'em up." He had no excess capital with which to operate.

The two men struck a deal that made Burdick & Burdick the exclusive dealers in the region for Challenge windmills and water-supply equipment. This was the big break that enabled Burdick & Burdick to become a major provider of water supply equipment in the desert Southwest. The start, however, was tenuous. Redman shipped a carload of windmills to El Paso and Burdick & Burdick paid the freight with the understanding that, in Redman's words, "Every time you put up a mill, get your money and send us a check." Keeping his end of the agreement, Tex relates, "every time I got a mill up, I ran down to the bank and got a cashier's check and sent it to 'em."

The windmills that the Challenge Company sent to Burdick & Burdick over the years were all the Challenge 27 model, which the company had introduced in 1927 and which it sold until just after World War II. They were modern self-oiling steel mills that were made in 6¾-, 8-, 9-, 10-, 12-, 14-, 16-, 18-, 20-, and 22-foot diameters. Burdick also sold Challenge Company pumps, prefabricated galvanized steel towers, and a full range of their auxiliary water-supply goods.

The dark days in the depths of the Great Depression did not offer the best prospects for the

fledgling business, but hard work and long hours, combined with the needs of ranchers for wind-pumped water, overcame some of the difficulties. Even so, Tex recalls, "It was hard times in those days. . . . We bargained for our mills. We bargained for the erection . . . . We would put up a mill for sacks of beans and things like that." The Burdick & Burdick firm managed to pay its bills and keep its books in the black, in the process gradually building its trade and list of satisfied customers who returned over the years.

By the mid-1930s, Burdick & Burdick was becoming known as an important windmill distributor. As Tex remembers, "It finally grew and grew until we hired another crew, and then we hired another crew." With several erection crews on the road all the time, he began trying to dodge back and forth between them, arriving about the time that an installation would be reaching its completion. "In other words," he explained, "I would like to be at a site when the mill was completed to clean up the mill, . . . to make the collection, and then to head over for another crew." It was a hectic time for Tex Burdick, but it built the business into a successful operation that moved from the tin barn into a newly constructed brick building at the same location on North Cotton about 1935. Continuing to grow each year, Burdick & Burdick celebrated its highest-volume year for windmill sales in 1941, when it marketed eight railway carloads of new windmills in a twelve-month period. By this time

B. H. "Tex" Burdick, Sr., March 1991, holding the No. 2-C Kodak Jr. camera with which he made the photographs in this book. Photograph by Warren Thwaites.

the firm had three crews in the field, each in its own truck. The high times for sales came to a halt when the nation entered World War II in December 1941.

Burdick & Burdick continued its business of selling, installing, and repairing water supply equipment during the war years, but the red tape of rationing and priorities cut deeply into the trade, as did the limited availabilities of certain items. Tex wrote to President Franklin D. Roosevelt every day for a month asking if there were not some way for firms like Burdick & Burdick to provide water-supply goods to ranchers without the inordinate delays of processing paperwork through the wartime ration boards.

During the conflict another event occurred that shaped the history of the firm: An insurance inspector called on Tex Burdick to investigate his liability insurance coverage. After his visit, the representative declared that Burdick's windmill erectors legitimately should be classified for insurance purposes as "riggers," as were comparable workers in the oil fields. This meant that Burdick's liability insurance rates for these employees would jump to eighteen dollars per hundred dollars of salary. As Tex relates, the firm "couldn't afford to pay eighteen dollars a hundred. We decided to abandon sales to the ranchers and go through dealers." Consequently in 1946 Tex Burdick sold most of his trucks and erecting gear and advised his erectors who had managed to stay with the firm through the war that they would be on their own, and Burdick & Burdick became a wholesale dealer of water supply goods. Although windmills were declining in sales nationally, Burdick was able to take up the slack for his company through promoting such lines as iron pipe, power pumps, pneumatic water supply systems, and stock-watering goods. The shift away from windmills was a major change, but it worked.

Always looking for new ideas, he decided to use a private airplane to give Burdick & Burdick increased visibility. A licensed pilot who owned a share in a small pleasure plane, he traveled in 1946 to Wayne, Michigan, to fly home solo in a bright red Stinson Station Wagon high-wing monoplane. Because his license did not permit him to use the plane commercially, he made it a public relations instrument for Burdick & Burdick.

On one occasion Tex received a late-night telephone call from a dairyman a hundred fifty miles away at Gila, New Mexico, northwest of Silver City. Lightning had struck his operation, burning out the motor that ran his Berkeley pump. The man was an old customer, and Burdick knew that the entire dairy and its cattle were dependent on the water from this one source. Asked for a quick replacement of the pump, Burdick replies, "Well, you meet me on some straight road up there between daylight and sunup." The dairyman must have breathed a sigh of relief; Tex was going to make the delivery himself. About three o'clock the next morning, B. H. Burdick got up, dressed, and drove to the store to pick up the needed motor. From there he headed for the airport, filed a flight plan, and took off to the northwest in the red Stinson Station Wagon, all in the dark. "I took off and got up to Gila . . . just as it was breaking day," he remembers. Having flown over the little town to alert the dairyman of his arrival, perhaps waking half the community, Burdick landed on a straight stretch of county road. The customer drove him to the dairy, where the two men installed the motor easily and returned the pump to service. The dairy herd had its water. Word spread in the Silver City area about Tex Burdick's emergency delivery. "With that kind of service," the businessman states proudly, "we got lots of repeat orders from that part of the country."

In a unique public relations gesture to his regular customers, Tex and his wife Juanita would take early Sunday morning flights along their own complimentary Sunday newspaper route. "If we were going west," he relates, "we'd stop at Bradesfoot, and we'd stop at the Johnson Ranch, and the Hyatt Ranch, and we'd go right on to other ranches." After circling each headquarters, they would drop the Sunday morning paper from the air as close to the house as possible. Sometimes they managed to drop them a little too close for comfort. One time Burdick hit a horse at a roundup, and on another occasion the Sunday paper passed through the front window of a house and landed in the owner's bedroom. Tex offered to pay for the repair, but the old customer east of El Paso smiled and said, "You just drop me the paper, and I'll pay for the window."

Realizing that times were changing in the water supply business, Tex foresaw that he would have to begin offering new lines of products. He decided "to take on plastic pipe." Ordinary business wisdom advised caution with a product as new as plastic pipe in the years following World War II, but Tex was convinced that it represented the strongest future for Burdick & Burdick. Questioned by his son, who each year was receiving a percentage of the business, Tex "told him that someday we'd be able to sell nothing but plastic pipe and still make a good living."

Time bore out Tex Burdick's judgment, for today Burdick & Burdick sells plastic pipe by the railway carload in the way that it once sold windmills. Tex Burdick retired from the enterprise in the 1970s leaving it in the hands of his son and grandsons, who today direct its activities from a handsome store at 1701 Myrtle Street. Burdick & Burdick is alive and well, still in the business of distributing water supply equipment. It even sells a few windmills each year, though the purchases are probably as much for sentimental reasons as any others. A quiet but profitable business enterprise, Burdick & Burdick through the second half of the twentieth century has been known to most of its customers as an old and reliable wholesale company dealing with pipe, tanks, pumps, and general water supply goods. Few people except for old customers and friends of Tex Burdick are aware of the historical significance of the firm as a purveyor of windmills.

In 1975, archivist Larry D. Sall, representing the Texas State Library and its Regional Historical Resources Depository system, learned of B. H. "Tex" Burdick, Sr., and his experiences with windmills and interviewed him on tape. Shortly thereafter, Sall informed this author of the interview and recommended that Burdick be interviewed more extensively. I made arrangements to take up with Mr. Burdick where Mr. Sall had stopped in his investigations. I discovered that Mr. Burdick had preserved a large and historically significant collection of his photographs of windmill work throughout the desert Southwest covering almost a quarter of a century. The images numbered almost four hundred, both snap-shot-sized prints and 3- by 5-inch nitrate based negatives. As interviewing progressed with the eighty-year-old Mr. Burdick, I began reproducing the entire collection of windmilling images, some of which were represented in both negatives and prints and some of which survived only as one or the other. This substantial undertaking in reproducing the historic photographs was conducted in 1979-81 under the auspices of the Panhandle-Plains Historical Museum in Canyon, Texas, where I was serving as curator of agriculture and technology. The copy negatives and a set of 5- by 7-inch prints are in the Research Center of the Panhandle-Plains Historical Museum, where they have been available to researchers for a decade. I have used a number of the reproductions to illustrate scholarly books and articles, and I organized exhibitions of enlargements of the images both for the Panhandle-Plains Historical Museum and for the Strecker Museum at Baylor University. In 1991 I resumed interviewing Mr. Burdick, combining this work with an interview with one of his former employees, Mr. Harry J. Clifford. This final stage of research was assisted with support provided by the University Research Committee of Baylor University.

B. H. "Tex" Burdick's hundreds of surviving photographs, taken from the 1920s to the 1940s primarily for purposes of public relations and sales promotion, represent today a singular collection of images recording a nearly vanished American vocational subculture, that of the professional windmill men. The subjects of these photographs were individuals whose lives centered on erecting and maintaining water-pumping windmills. Although similar windmillers once operated throughout virtually all of the American West, no large groups of crews like Burdick's in the 1930s remain anywhere in the United States today, and no other comparable collection of images showing all the phases of windmill work is known to exist. The photographs record virtually every aspect of historic windmill work, from the excavation of anchor holes to the antics of gleeful erectors balancing themselves atop the hoods of completed windmills high in the air.

Through Tex Burdick's photographs, windmills are no longer mere factory-made galvanized structures standing inanimately above the countryside. They become steel monuments to men like Tony Venagas, whose bride wanted to accompany him into the field; Harry Clifford, daredevil at heights; and Sid Bowlin, master of the slumgullion stew. Real people like W. R. Lovelace and John Prather had thirsty livestock that needed the water pumped by windmills, and other real people like Ramon Delgado and Bill

Berry tore their hands and mashed their fingers to meet that need. These photographs demonstrate that behind every windmill in the field there were and may still be actual people who built and used them.

Tex Burdick did not think about creating a historic record when he photographed his men at work in the field. As Harry Clifford remembers, "Tex always had his camera with him, taking pictures of this, that, and the other." From photographing mountain scenery when he worked at Hanover, New Mexico, Burdick shifted his focus in El Paso to his homemade well drilling rig and to windmill jobs that his father arranged for him. After Tex began erecting windmills on his own in substantial numbers, he decided to start recording his work so that he might show the pictures to prospective customers. "After I completed a job, I thought I'd better take a picture of it," he said, adding, "I very religiously . . . saved these photographs and put 'em in a book." There he began pasting typewritten captions to identify customers, locations, depths of wells, and sizes of pipes and pumping cylinders. "But, after a while I got to putting up so many of them," he admits, "I didn't finish up." He did, however, continue to throw his No. 2-C Kodak Jr. camera in its leather case into the front of the truck, carrying it to job sites and photographing the men at work. Time did not permit Burdick to keep up his careful captioning, but he did continue making the photographs.

When asked about his motives in producing an extraordinary collection of documentary photographs, Tex Burdick remarks only, "I just wanted to get as good a picture as I could. For the result." For Tex that result was the actual installation in the field. The creative act for him was not the photograph—it was the actual functioning water system.

"It always gave me a thrill to go out to a well," Tex explains. There he would find a piece of pipe sticking out of the ground perhaps a foot or more, and on the surface would be the tailings where the driller had dumped his slush on the ground. For Burdick the thrill was, as he says, "to go out there with a truck and a trailer and put up a mill and tower, run the pipes and the rods, . . . put up a tank, . . . put concrete in the bottom, . . . and the outlet pipe." Burdick's emotional high came at the completion of the job. "When you got all that together and water coming out of the pipe . . . and hitting the bottom of that tank, it always gave you the feeling of accomplishment." As the members of the crew reached that point, they could stand up, look over their work, and think to themselves, "That helped somebody out."

Tex Burdick concludes, "That was a good feeling when you did that."

The Photographs

Lee Burdick climbs the ladder of a Standard wooden-wheel windmill. B. H. "Tex" Burdick's Model T speedster is parked near the base of the tower. Circa 1923.

The early 1920s found Tex Burdick in El Paso and looking for work. "My father . . . was running a cotton gin," Tex remembers. "He was free after about the first of the year, when the cotton was completely ginned, until the repair of the gin, which would start again about the first of July." This annual schedule left Lee Burdick free about half the year to job hardware and supplies to his old customers, ranchers throughout the desert Southwest. Tex continues, "He . . . would go out to these ranches (and he knew every rancher) and sell either barbed wire or pipe or equipment."

While Lee Burdick was jobbing hardware to the ranchers, he also was busy drumming up business for his unemployed son. "He would make the rounds for me to pull the wells or to releather the well or to put up a windmill, and in that way he and I began working together," Tex notes. "We used to build 'em on the ground and raise 'em all in one piece, the wheel and tail and everything on 'em. . . . I would do the erections, and he was doing the selling."

Tex Burdick's water-well-drilling rig. El Paso Valley, late 1920s.

Tex Burdick earned some of his first "real money" with a homemade well-drilling rig, but at the beginning he didn't know a thing about drilling wells. "After I got married, . . . there was a fellow by the name of Jack Helms who was a rancher's son," he relates. "Jack Helms and I decided to put in wells."

The rancher's son was already superficially familiar with well drilling, but Tex was not satisfied that the two of them knew enough to do it successfully. The city of El Paso was putting in a big water well with a rotary drilling rig, so the two young men decided to take advantage of the situation. "Jack and I would go down there each morning when they opened up, and pretty soon we were working down there with them. We weren't getting any pay, but we were learning," Burdick says.

After two months as volunteers for the city, Tex and Jack decided that they were ready to become well drillers on their own. Their first step was to build their own equipment. "There was an old Mack truck with solid rubber tires on it. It didn't have an engine or anything else, just a chassis." Using it as their framework, Burdick and Helms put together their own shop-made jetting-type well rig, the motive power coming from a cast-off Overland automobile engine.

The portable rig employed weights that struck the steel pipe casing to drive it into the ground, while a smaller-diameter pipe on the inside flushed water under pressure from the base of the casing. The water loosened the sand, clay, and gravel and carried the debris to the surface. As Burdick explains, "The weights would hit on top of a tee on the top of the drill pipe and drive the pipe down, and the cuttings would come out between the inside drill pipe and the casing."

Before long Jack Helms lost interest in drilling wells, but Tex continued in the business for several years. He focused his activity in the El Paso Valley, where the jetting rig was effective on the fairly fine-grained, unconsolidated subsurface materials. "I don't know where we got the idea," he remembers. "We got it and made a lot of money on it. We could put down a well, and believe it or not, we put down the well for $1.35 a foot and furnished the pipe . . . and made a profit, made good money."

26

Burdick & Burdick windmillers Ramon Fernandez, Carl Boyd, and Sid Bowlin (from left) with a load of flat steel sheets for tank construction in the alley beside the store at 190 North Cotton Avenue. El Paso, circa 1938.

Among the services that Burdick & Burdick provided to its customers was the construction of large, open-topped, circular steel tanks for the storage of water for livestock. When the wind failed to blow for several days, such tanks could hold enough water for the livestock until the calm ended and the windmill returned to pumping. Burdick crews built these tanks from second-hand six- by twenty-foot steel sheets which had been used before as components in oil-field storage tanks.

Speaking of the large steel sheets, Tex Burdick recalls, "I bought those from W. Silver Company . . . there in El Paso. . . . They had holes punched along all sides." The big sheets were not necessarily clean or attractive. "Sometimes these would have tar on the inside," Tex says. "I would take one of these cars and hook onto them at Silver's . . . and pull them over with that tar down." Occasionally they required special efforts at cleaning just so that they wouldn't get everything else in the trucks dirty with black tar. "I'd pull 'em right along the street," Burdick recalls, and "turn the corners fast to pull a lot of the tar off."

28

29

Loading the big GMC truck with two disassembled Challenge 27 windmills. El Paso, circa 1941.

A windmill crew began an installation job by loading tools, supplies, and windmill pieces onto the truck or trailer outside the Burdick & Burdick building on North Cotton Avenue. Former employee Harry J. Clifford remembers that the men would get things ready at the big front and back doors and then "back this truck up and load it on. . . . We'd get everything . . . in front . . . and go 'round to the back and load our chuck box and our bedrolls and other stuff."

The back bumper frequently held the steel cables used in raising towers or lifting heavy components. As Tex explains, "It was easier than trying to roll it up on a reel," and it kept the cable from twisting. Removing the cable from the bumper was easy, for one man simply drove the truck forward slowly while another walking along behind the vehicle played it out onto the ground. From there the men pulled the cable through the blocks and tackle and readied it for use.

Harry Clifford notes that "the casting was the heaviest part of a windmill." Also known as the head, the main casting generally was loaded first because of its awkwardness and weight. "It tied to the . . . back of the cab," Clifford remembers, "chained down with boomers." Moving such parts was no job for pantywaists. "They had a hoist in the back of the building that would put the big ones on," he says, but "the little ones, well, anything up to five hundred pounds, we usually loaded 'em by hand."

31

Andres Avila poses behind the head of a twenty-two-foot Challenge 27 windmill in the shop of Burdick & Burdick on North Cotton Avenue. El Paso, circa 1940.

Andres Avila, known to the windmill men as "Nechi," was the shop man. He came into the North Cotton Avenue building at seven-thirty each morning to clean things up for the new day and then retreated to his bench to work on pumps in need of repair.

Many local people knew Avila best as the fresh meat man, because Tex Burdick on return from the field often brought him stacks of freshly killed rabbits. As Harry Clifford recalls, "Tex had a longbarrel .22 pistol, and he was really good with that thing." Time after time, the boss would stop the car or truck, take aim with the handgun, and shoot a rabbit. "He'd never get out of the car," Clifford says, "and we'd get the jackrabbits and put them in the trailer and keep going."

Old Avila lived the next street below Burdick & Burdick just a short distance from downtown, and his yard was fenced with pickets. "I'd . . . pull up with the trailer," Burdick says, "and the rest of the boys and I'd take these jackrabbits and hang their heads in this fence." As soon as Avila's neighbors saw the windmill men stop by his house, "they'd all come, and they'd want these jackrabbits," according to Burdick. "They'd

want *carne fresca* . . . fresh meat." Avila would meet them at the fence and start giving away rabbits. Nothing pleased him more than to dispense the fresh meat, saving the cottontails for his own family.

One of Avila's sons served in the Pacific during World War II. After participating in the capture of one of the enemy-held islands, he was among a large party of American soldiers who decided to camp in a level area at the top of the island, apparently oblivious to substantial numbers of Japanese troops at large. Tex Burdick remembers, "The Japs blew it up and killed I don't know how many . . . and his son was one of the boys that died in that . . . according to the article that they sent to him."

Andres Avila worked for Burdick & Burdick so long that most people saw him as a permanent fixture in the store. "He worked for us . . . I guess eighteen, nineteen, twenty years," Tex says. "And we retired him, paid him a hundred dollars a month after we retired him. And in those days that was pretty good pay."

32

Burdick & Burdick employee Ramon Fernandez stands in front of the overheated 1928 or 1929 Ford Model A truck loaded heavily with parts for a twenty-foot Challenge 27 windmill circa 1935.

Preparing to leave for the field, Burdick & Burdick employees carried with them everything that they might need, for many of the installation sites were in truly remote country. All the big trucks were fitted with auxiliary fuel and water tanks just for such use.

Recalling the preparations for departure, Harry Clifford recounts, "We always had our digging equipment: picks and shovels. . . . The windmill . . . casting was loose. The vanes and stuff were all in the crates. The tail was loose. And the angles for the tower were all bundled up in bundles of four."

For Tex Burdick the ideal windmilling truck had one- to two-ton capacity and a flat bed. He always installed an inverted U-shaped steel "headache bar" just behind the cab to carry steel sheets for tanks, to transport assembled sections of towers, or to hold down heavy main castings or other big loads.

By the time Harry Clifford had become a Burdick employee in 1936, the Model A Ford in the photograph was already a high-mileage vehicle. "We had a lot of trouble with it," he remembers. "It was getting old when I went to work for him, but it'd go out anywhere. We'd load it up and take off . . . usually three guys in the cab. . . . It went everywhere we wanted it to go."

Burdick considered the Model A an ideal windmilling truck, in part because of its almost unlimited carrying capacity. "We put on all we could get on. . . . They had to get there. . . . That's all there was to it. We never knew anything about overloading. Just as long as it would go." He concludes, "That Model A there was a darned good truck."

34

An unidentified Mexican crew member poses beside a steel tower with an eight-foot Challenge 27 windmill erected for Dr. J. D. Love. Hatch, New Mexico, circa 1935.

When Tex Burdick hired men for his windmilling crews, he didn't necessarily seek the traits that one might expect. "We didn't choose the boys because the were adept at crawling up on a tower," he states. "We chose them because they could dig anchor holes and mix cement. That was the most important." Burdick wanted muscles and discipline.

The typical crew had three men, four for erecting the largest of the mills, with one in charge called the top man. He was the one who worked most atop the towers, the others assisting in whatever was appropriate at the particular stage of erection. The laborers on the crews, for example, laid out the angle steel, girts, and braces for steel towers in their proper sequence so that they could be assembled efficiently. According to Burdick, "They became very adept in doing that. After three or four or five or ten or fifteen mills, they could lay it out all to themselves."

Burdick & Burdick crews were always bilingual, the Anglo supervisors speaking Spanish and the Hispanic laborers communicating in English. Harry Clifford remembers, "We had no language problems at all." He had begun speaking Spanish as a small child growing up on an El Paso Valley dairy where he was "out bumming tortillas and beans" from mothers of Mexican families living on the same property. Tex Burdick, in contrast, spoke Spanish that he learned at El Paso High School, and his Hispanic employees sometimes snickered at his Spanish, although they had no difficulties communicating with him.

Only one of the Mexican employees, Ramon Delgado, ever became a top man who headed a work crew. He became "the man that went out and put 'em up," Tex Burdick recalls. Delgado was unmarried when he went to work for Burdick & Burdick, but before long he decide to wed. Tex Burdick relates that the young bridegroom with the boss's permission took his pretty wife out into the field, where in camping he placed their bedrolls some distance from the others. "She'd never been out and slept on the ground. It was a scream. She wanted to sleep in the truck," Tex reports. He happened to visit the camp on the bride's first trip out, noting that by the time he arrived on the scene "she was just about to quit him."

To settle the squabble, Burdick offered to the new bride and groom, "What I'll do, next time you come in, . . . we'll take your truck and put a tent on it." The new bride continued to accompany Delgado to the field for two or three months, but after that she decided to stay at home and leave the camping to the men. Besides, Burdick adds, "she was in the family way by then."

A twenty-two-and-one-half-foot railroad-pattern Standard windmill pumps water for livestock. W. N. Fleck Ranch south of Alamogordo, New Mexico, circa 1923.

The Burdick crews worked on some truly magnificent wind machines, though the men in the field rarely viewed them that way. The huge Railroad Eclipse mills were the largest of the Eclipse line made by Fairbanks, Morse and Company. With wheels and vanes made from wood, the Eclipse mills were the most common in the West before the turn of the century.

Invented in 1867 by a missionary among the Ojibway Indians of Wisconsin, within two decades, the Eclipse came to its prominence in the trans-Mississippi West. In 1901 the rights to its patent entered public domain, and almost immediately several companies began producing virtually identical mills with interchangeable parts in bids for portions of the Eclipse market. One of these Eclipse copies was the Standard made by the F. W. Axtell Manufacturing Company in Fort Worth, Texas.

Referring to the Eclipse and Standard mills, both common in the Southwest half a century ago, Tex Burdick declares that neither was necessarily better than the other. Because of the relative proximity of the Fort Worth factory to El Paso, "The only advantage we had was because the Standard was made down here in Texas," says businessman Burdick, "and we could get the repairs and everything else real quick."

The mill in the photograph pumped water on the W. R. Fleck Ranch south of Alamogordo, New Mexico, and poor quality water it was. "This water . . . had so much sulphur in it," declares Burdick, "that they had to aerate it before the cattle would drink it."

Burdick & Burdick would undertake repairs to the old wooden-wheel windmills, but more frequently they were called upon to replace them with modern Challenge 27 steel mills. When the tower was to be reused, the men disassembled the old mill so as not to injure the derrick. If the towers were not to be used, however, the crews took them down the fastest way they knew. "We'd take a cable and tie it to this old truck," Harry Clifford remembers, "and just start off." That way when it came time to cook meals, "we did not have to cut any kindling . . ."

38

From left: a back-geared, self-lubricating Challenge 27 steel windmill, a sixteen-foot Railroad Eclipse, and a fourteen-foot regular pattern Eclipse windmill on a ranch in the desert Southwest. Circa 1932.

There were two major distinctions between the old wooden windmills and the newer steel mills besides the obvious difference in their material of construction. One related to maintenance, and the other to performance.

The modern steel windmill manufactured since the 1910s and 1920s have been virtually all self-lubricating. Using technology introduced to the market about 1912 by the Elgin Windmill Company, these mills have working parts that operate in baths of oil comparable to the crankcases of automobile engines. The internal moving parts are flooded with oil and are covered with metal hoods for protection from the elements.

The self-lubricating feature, introduced to the Challenge line in the 1920s, meant that the new machines did not have to be lubricated weekly by hand, as was the case with the earlier mills like the Eclipse, which had all their working parts exposed to the weather. The older-style mill demanded regular lubrication, requiring weekly climbing of the towers, a chore many owners and cowhands detested. Harry Clifford disparaged the acrophobic ranchers, "I've seen 'em take ten-foot pieces of pipe, tie an oil can to it, and reach up and oil the exposed gears. They wasn't going to get up there at all."

The second major feature of the newer steel mills, an idea that appeared on older mills with exposed working parts as early as the 1880s, was their back-geared design. With the addition of gears, mills with wheels having more efficient curved steel blades were able to make several revolutions for each stroke of the pump. Operating in much lighter wind, usually beginning to turn when it reached about six miles per hour, they worked more hours of the day, and, although they raised less water in strong winds, they provided considerably more water in the long run.

Even though the newer back-geared, self-lubricating steel mills required far less maintenance and pumped more water, many of the traditional ranchers in the Southwest preferred for years to stick with the older direct-stroke wooden mills. Tex Burdick understood their logic, in spite of his company sales of the more modern mills.

"When you want to get water and the wind is just right," he admits, "a direct-stroke mill will pump more water than any back-geared mills," repeating the very important qualification, "when it's just right." He advised his customers correctly, however, that the back-geared mill continued to operate when the winds were too light for the less efficient wooden mill.

Harry Clifford describes the direct-stroke mill succinctly: "They pumped water every time they went around, but they didn't go around so often as the others."

41

A Burdick & Burdick crew arrives with a twenty-foot Challenge 27 windmill on a New Mexico ranch. Circa 1935.

The arrival of a Burdick & Burdick crew to install a new mill at a ranch was always an occasion of some interest to the cowhands. They knew that their boss was spending a considerable sum for the windmill and perhaps a new tower—more money than they might earn in several years—so they all tried to come in and take a look at the shiny galvanized steel and freshly painted iron that represented such an investment in future water.

"They'd usually come out and look it over and walk around it," Tex remembers. While the cowboys peered into the truck or trailer and speculated on the use of the various parts and tools, the top man of the crew would seek out the ranch owner or manager to inquire about the specific site of the new mill.

"I'd ask him where the well was that I was going to put this thing up," Burdick says, "'cause very, very seldom did we ever put one up at a headquarters. It was always out some place six, eight, ten, twelve miles. And then we'd be out there all by ourselves." Occasionally a ranch employee would lead the way to the site (often marked only by a piece of well casing protruding from the ground as it had been left by the driller), but usually someone just told the crew where to find it.

The important thing for the rancher was for the windmill men to get to the site and start work without delay. "They wanted us to get it up so their cows could have some water," Tex concludes.

43

Reclining on a pile of tools and gear, top man Carl Boyd rests after the completion of an installation. 1942.

In the late 1930s, Carl Boyd became one of B. H. Burdick's top men, heading an erection crew and performing the skilled work atop the towers. Of all his crew chiefs, Tex Burdick relates the most vivid memories of Boyd, a man of mystery from Georgia who would reveal no details of his earlier life. "When I hired him," Tex says, "he'd been working for a candy company, . . . and I told him that the work was hard." As a new employee, Boyd ended his first day as a windmiller with his hands covered in blisters. "On the second day these blisters broke, but he kept on working," his boss remembers. "He said he wasn't going to work for a candy company all his life, and he was going to get out and see some of the country." Finally, as Burdick says, "the blisters began healing up, and he began hardening up." Boyd worked for the company until 1946, when it phased out windmill erecting.

Boyd took to windmilling the way ducks take to water. "He really enjoyed it," his boss remembers. "He said there was something motivating about working on them. . . . He liked taking a crew out and going out on these jobs."

During most of the Depression years, Boyd had no home of his own. He camped out while in the field, and in El Paso he simply bunked down on his bedroll at the Burdick & Burdick warehouse. Fellow employee Harry Clifford recalls, "He slept at the shop and ate around there at that little cafe," the Rock Hut Cafe at 1612 Texas Avenue, patronized by many members of Burdick's crews.

"He was a hard worker," B. H. Burdick states. And adds, "He was one heck of a nice guy."

Tony Venagas, Harry Clifford, and Carl Boyd (from left) with bird dog Jerry in camp. Warner Ranch north of Rodeo, New Mexico, December 1941.

Windmillers' camps were not always the neatest places, especially when the wind disturbed everything that was not tied down. "It looks like we were trashy," comments Harry Clifford, pictured seated at the center of a group of windmill men.

This particular camp held special memories for all of the men who stayed there, memories not necessarily related to the job that they did. On this occasion Tex Burdick had a terrible cold, his nose running constantly. "The wind was blowing, and it was just as cold as the very dickens," he said. Packets of soft paper tissue were newly on the market. "I had a box of Kleenex," he notes. "About every two or three minutes I'd grab that Kleenex and then turn it loose, and it would go out there a mile and a half . . . I'll never forget that."

Only Burdick remembers the tissues in the wind, but everyone on the crew long remembered what they heard on Sunday evening, December 7, 1941, on the car radio. As Burdick relates, "I was on this job and listening to the radio that night when I heard about Pearl Harbor." All

through the night the men sat in the car listening to the reports of the Japanese attack on the U. S. Pacific fleet in Hawaii. "Then all of a sudden," Harry Clifford says, "about midnight these airplanes started coming over." The windmillers didn't know where they had come from or where they were going, but the effect was electrifying. "I just wondered how long it would be before I got in," Clifford recalls thinking as he sat in the desert in the middle of nowhere listening to droning aircraft engines high above. "We didn't know whether we ought to quit and go to sign up or wait till they came and got us."

They strained that long night to hear the words over the air. Burdick remembers, "I left the radio on in the car, and the next day we had to push it to get it started."

Top man Carl Boyd (right) with an unidentified company employee enjoying a meal eaten from the vane sheet of a Challenge 27 windmill as bird dog Jerry watches for scraps. Southern New Mexico, 1942.

"I worked for Tex Burdick . . . for two dollars a day and all I could eat," states former employee Harry Clifford, and indeed one of the attractions of employment for Burdick & Burdick was food. All through the depths of the Depression, Burdick paid his laborers fourteen dollars a week plus all their meals when they were on a job. "We usually ate pretty good," Clifford comments.

Meals at erection sites were full of fat and calories, for the men needed all the energy they could muster in order to maintain their arduous labor day after day. "For breakfast we'd have scrambled eggs or fried eggs and potatoes, usually toast," Clifford notes. His boss adds that sometimes he provided bacon, but not too often. Clifford states, "Maybe we'd make pancakes in the Dutch ovens."

A regular item of fare in the windmillers' camp was coffee—almost from the moment of arrival in camp. "We always . . . had a pot of coffee on," Burdick said, "Just as soon as they got up they had it." The coffee itself came from Juarez across the Rio Grande "because the Mexicans liked it."

Lunch, according to Harry Clifford, usually consisted of sandwiches or maybe a big bowl of stew that had been simmering on the campfire. Dinner generally was a heavier meal, frequently steak and potatoes. Occasionally the ranchers would give the windmill crews fresh meat like beef or mutton; if time permitted, the men themselves might hunt for rabbits. Perhaps the most important staple in the diet, however, was pinto beans cooked in a big pot that usually was bubbling on the fire.

Windmiller Clifford observes of the meal preparations, "We all just pitched in and did the cooking." The fare must have been pretty good, for Clifford recalls with a smile, "We never throwed much out."

49

A Burdick & Burdick laborer excavates an anchor hole for a windmill tower at an erection site. Circa 1935.

For men on the Burdick & Burdick crews, digging anchor holes for the towers was the most disagreeable part of the job. Harry Clifford declares the digging was "the hardest part of windmilling," and he was not alone in the feeling.

Burdick as the boss handled the excavations as fairly as possible. "If there was four of us," he says, "each one of us dug our own anchor hole . . . and the four of us, why, nobody helped anybody else." Windmiller Clifford agrees that "everybody dug his own. . . . If there was two men, you dug two and two."

The excavations for tower anchor posts were not like simple postholes. They were large, far larger than might be expected. "We learned from experience," Burdick explains, "that a hole about two foot wide and four foot long . . . gave you the chance to get your shovel down in there." The excavator alternately dug on one side, threw out the dirt, and then turned around to work on the other side, many of the men taking pride in the careful rectangular shapes they cut as they sank their pits. Remarking on the dimensions, Burdick notes, "It had to be big enough so you could get down in there."

Not only did the holes have to be five feet deep (deeper for the very largest towers), but they also had to allow precise alignment of the anchor posts that supported the towers. "If you were an inch off at the bottom," Clifford notes, "two legs would be an inch higher than the others" at the top, and they would not fit together properly. If this occurred, he said, "you'd have to saw 'em off or bend 'em or twist 'em, so we were very careful about getting them level and plumb at the bottom."

Indeed, digging anchor holes was the worst part of windmilling. Tex Burdick himself admits that it was "just a lot of darned hard work."

51

A Burdick & Burdick crew poses on a just-completed wooden tower with "sprung" legs. New Mexico, circa 1935.

The first windmill towers were built of wood, and although steel had been the predominant construction material since the 1890s, many of B. H. Burdick's customers still preferred the traditional wooden towers. In the arid environment of the desert Southwest they lasted almost as long as steel. Generally the landowners provided the material for the wooden towers, and the crews cut it to size and assembled the pieces into derricks. "I didn't like 'em," Harry Clifford declares of the wooden towers. Their construction required far more work than the prefabricated steel towers.

The men first assembled two opposite sides of a wooden tower. Then, with one side stacked on top of the other, they spread apart the lower end of the structure and installed the horizontal girts and diagonal braces to form the remaining two sides. All of this work was done before the tower was raised.

In the best of the wooden towers, those termed "sprung," the lower legs were forced slightly out-ward, giving them a somewhat bowed appearance and placing the entire structure in tension to increase its rigidity. In this procedure the top of the tower was fastened together first. Then as the crew worked its way downward, it used jacks to force the four legs apart and create the desired tension. "You have to be careful," Burdick advises, "that one won't be a little more pliable than the other." It was easy, in other words, for the erector to spring one leg out more than the others, because no two timbers would have precisely the same tensile strength.

The crew would cut the four horizontal girts for each stage of the tower at once, making certain that they were all the same length (allowing for overlap at their ends). "Cut them all the same time and then make the tower fit the girts," Tex Burdick advises. "If you go to measuring and then just cutting, it never comes out right at the bottom."

Once the wooden tower was assembled and its anchor posts were secured, it was time to raise it into position.

53

A Burdick & Burdick crew erecting a fifty-foot steel tower for a Challenge windmill. Carrizozo vicinity of New Mexico, circa 1935.

The Southwestern ranchers who hesitated to give up their wooden towers in favor of the mass-produced galvanized steel ones knew that the old towers built of heavy timbers had served well for many years. They had a hard time understanding how the comparatively flimsy-looking derricks made of angle steel could stand up to the tremendous stresses and strains brought by windstorms. Little did they know the engineering that had gone into the new design.

"One of the hardest things," B. H. Burdick states, "was to get the ranchers to change over from wood to steel." Finally he proposed to one rancher, "Now, I'll put you up a steel tower, and if you don't like it in a year, I'll take it down and put you up a wood tower." The stockman still wasn't convinced. "Now, look at oil wells," Burdick argued, pointing out the change in oil-field derricks from wood to steel construction. Finally the stock raiser consented to the trial, and at the end of the year he was pleased with the steel structure.

"And all of a sudden," Burdick remembers, "they all found out that it was easier, quicker, and safer to put up a steel tower than it was to put up a wooden tower." Steel towers took over in the range country.

A Burdick & Burdick crew raises a sixteen-foot Challenge 27 on a thirty-foot steel tower for Dr. Francis Cole. West of Las Cruces, New Mexico, circa 1931-1932.

After a windmill tower, either wooden or steel, was assembled and its anchors were installed, the windmillers raised it to its vertical position. This could be the most ticklish stage in an installation.

First the crew used jacks to elevate the top of the tower three to five feet, blocking it up temporarily. "When we knew we were going to do this," remembers Harry Clifford, "we'd take the old windmill crate . . . and nail it back together and put it under the tower." Safely at ground level, the men would assemble the windmill proper on the top of the tower. On smaller mills even the wheel was put together at this stage, and occasionally even the wooden pump rod was installed, running from the mill downward through the tower. Now the windmill men were ready to raise everything.

In simplest terms, the raising of a windmill was an operation in which the crew members tied a steel cable to the top of the tower and pulled it up with a truck or a team of animals. In actual practice, however, the job was more complex. A steel cable connected the top of the tower with a set of blocks and tackle. One end of the cable, passing through the pulleys in the blocks, attached to a secure anchor, while the other end of the cable fastened to the truck or the harness of the team. Guy lines on two sides and the rear secured the tower from tipping over the wrong way as it went up. To provide sufficient leverage for lifting the upper end of the tower, the cable between the tower top and the blocks and tackle passed over a vertical support, such as a pair of crossed pipes or timbers.

Once all was prepared, the driver pulled forward very slowly while the head of the crew checked over all the connections, supports, lines, and tower components. If everything was in order, he directed the raising to proceed, the key to success being a smooth, slow movement that avoided any jerking motion that might loosen elements of the rig. As soon as the tower reached its vertical position, the crew members pried the legs into place and bolted them to the anchor posts, and the tower could be considered secured.

A heavy-duty steel tower with double diagonal braces and a large opening in one side for convenience in well service is built up from the ground. Hatchet Mountains of southwestern New Mexico, circa 1934.

The crews did not always assemble the steel towers on the ground and then raise them into position. Some circumstances dictated that they instead build the towers up from the ground one piece at a time—in locations where there was no room to assemble the tower horizontally, for instance, or where a tower was too large to be raised in one piece without buckling.

During his years with Tex Burdick, windmiller Harry Clifford assembled many towers from the ground up, and he described the procedure in some detail: The first step was the placement of secure anchor posts. As Clifford remembers with distaste, "You go out and dig these four holes, which everybody hated, and get the bottom angles in." Then the crew connected the tops of the anchor posts with lightweight angle-steel horizontal girts.

The next step was to bolt on the first section of ten-foot-long angle steel tower legs so that they "stuck up in the air." Five feet up the builder attached a second series of light angle-steel horizontal girts. "And then you'd crawl up on that band before it was bolted in tight," Clifford continues, "and put the top band in." That phase of the installation placed the top girt roughly ten feet above the ground. Then the erector attached the diagonal steel braces that were mounted every ten feet and tightened all the nuts and bolts. With the attachment of the light steel ladder on one side, the first ten-foot segment was completed.

To begin construction of the next ten-foot section, the erector made himself a temporary seat with a piece of lumber from the packing crates diagonally across the inside of a corner of the newly installed horizontal girts ten feet above the ground. He started bolting on another set of four ten-foot-long angle steel leg segments, which he lifted from the ground with a rope. When the time came for the next set of horizontal girts, five feet above the worker's position "you'd pull it up and you'd balance it with your head, and hold it against the other angle and stick a bolt through right quick and try to get a nut on it." After securing the girts to all four corner posts, the worker climbed up five more feet and repeated the procedure, continuing in that way to the top of the tower in successive ten-foot segments.

With the tower up, the next step in erecting the windmill from the ground up was the assembly of the mill itself.

58

A Burdick & Burdick crew uses the four-inch steel gin pole to raise the head for a twenty-two-foot Challenge 27 windmill to the top of a fifty-foot tower. 1940s.

On a small mill, one having a diameter of eight feet or less, crew members manhandled parts to the tower top using ropes, but on a larger mill a gin pole was used. Usually guyed to keep it nearly vertical, the gin pole supported the blocks and tackle that hoisted the heavy windmill components during the installation or repair.

The windmillers chained the gin pole to the top of the tower. Its base rested on the platform and often was supported from beneath, while guy lines or a chain around the top of the tower held it upright. For a medium-sized windmill, a Burdick & Burdick crew would use a gin pole made from a piece of two-and-one-half-inch pipe, but a truly large mill called for a special four-inch steel pipe twenty-one feet long. Using blocks and tackle mounted on the gin pole, the erectors raised the main casting to the top. Once the main casting was installed, they raised the vane assembly and then the sections comprising the wheel. "You'd pull the windmill sections up one at a time," remembers Harry Clifford, "and in a big wind they'd float out there like a kite."

Installing the twenty-one-foot gin pole at the top of the tower was no small task in itself. Tex Burdick always participated in these major jobs, and he was the one who would make his way up the nearly vertical gin pole to hook the blocks and tackle into its top. "I'm the ape that always climbed that gin pole and hooked that three-block tackle in the top," he declares proudly. To accomplish the feat, he would first climb the tower, a little over ten feet up the height of the gin pole. Perched there, he would reach up and across to fasten a special chain vice to the pole. Then about three feet above the chain vice he would attach a thirty-six-inch chain tong. Fastened to the pole, these clamp-like tools created a pair of improvised steps about a yard apart that he would use as foot- and handholds as he swung from the top of the tower onto the gin pole and climbed upward toward its apex.

After clambering up the gin pole from the apex of the tower and while standing on the upper of the two improvised footholds, Burdick would hook the upper block from the blocks and tackle into the top of the gin pole. When he reached up to do this, he lifted far more than just the block with three pulleys, for added to it was the weight of all the steel cable suspended downward from it. "When you have fifty, sixty feet of steel cable hanging down," Burdick explains, "it's quite a job to put that three-block in the top, but I'm the ape that always climbed it. I never asked any of the boys ever to do that."

60

61

Windmillers demonstrate their agility during the erection of an eighteen-foot Challenge 27 on a forty-foot tower. Circa 1935.

"I never had a safety belt, never had a hard hat in my life," responds Tex Burdick to a question about safety among his windmilling crews, but he is quick to add, "We always looked out for each other."

On the Burdick crews whenever someone called out the warning "headache!" everyone stood absolutely still, for someone on the tower had dropped a tool or something else but knew where it was going to land. In contrast, when the alarm "run!" rang out, everyone dashed for cover, for, as Tex Burdick says, "a wrench slipped or was kicked off or something like that, and you saw it was going to hit somebody."

There were comparatively few stated safety rules for members of Burdick & Burdick windmilling crews. Although no one ever intentionally endangered any fellow workers, a number of men did things that clearly threatened their own lives and limbs if not those of others. Harry Clifford, for instance, never hesitated to walk out on the vane stems of large mills: "I would turn loose on these big ones, where it's two-and-a-half-, three-inch angle [steel] out to the tail maybe ten feet. I would walk out that tail and grab the tail and stand there."

Often when Clifford saw someone on the ground with a camera, he would call out or wave to them to "have my picture taken . . . standing sixty or seventy feet off the ground." That was the case in the photograph seen here. Tex Burdick relates that he was preparing to snap the picture when the man on the vane stem sang out, "'Wait a minute!' and he just walked out there." Explaining such behavior, Burdick says, "You get used to it. It's no more than walking across the floor."

62

Windmiller Harry J. Clifford with arms upraised stands balanced atop an eighty-foot tank tower with ten-foot Challenge 27 windmill. Warner Ranch near Rodeo, New Mexico, December 1941.

Harry Clifford, the man atop the tower with arms upraised, was truly one of B. H. Burdick's daredevils. Heights had no effect on him, so he did whatever was necessary on his job and had no second thoughts. Speaking of Clifford and the others, Tex declares, "The boys were just immune to height or anything like that."

When Burdick was photographing this newly installed tank tower and mill, Harry Clifford climbed atop the head of the machine because, as he explains, "I just wanted to get up on top of it and wave at 'em." Somewhat irked by the worker's intrusion into this public-relations photograph of the completed installation, Burdick grumbles, "He just wanted to be funny . . . just showing off."

Of the remarkable agility of his men, Burdick observes that they could "run up and down those towers just like a monkey on a string."

The statement was not exaggerated, for Harry Clifford resorted only rarely to a ladder when coming down a steel tower. Instead he placed his gloved hand inside the angle steel of a tower leg, crooked his right foot gently around its outer edge to serve as a guide, and began slipping down the outside of the leg, catching himself with his left foot on a horizontal girt every five feet just long enough to slow his fall to a controlled speed. Detailing his method, Clifford notes, "Each time [I] just take it [the left foot] off and go down and get the next one, and go down and get the next one. Slide right down with your gloves in the angle." From the water tank at the fifty-foot level on this tower, Clifford asserts, "I'd get down from there in less than ten seconds." Reflecting on his short cut to the ground, he opines, "It's a wonder I hadn't broke my neck."

An eighteen-foot Challenge 27 windmill on a wooden tower. Reaves and Manning Ranch, twenty-seven miles north of El Paso in New Mexico, circa 1935.

Blowing against a windmill and tower, wind-borne sand can create a surprisingly strong static charge. Longtime windmill men observe caution in approaching towers after sandstorms, but new-comers seem to learn this lesson the hard way.

"We went out to service a mill that we'd put up several years before," recollects Tex Burdick. The mill, a big steel Challenge 27, had a metal pull-out wire that was used to turn it on and off. The wire dangled from the mill atop the tower to a hand winch mounted on one of the tower legs near the ground.

In an effort to look eager for work, a new employee "fell out of the truck and ran over there to turn the mill out of the wind" by cranking the winch. He did not suspect the static electric charge, created by sand blown against the steel windmill, that had been conducted by the pull-out wire to the winch. "When he got up to within about three, four inches," Burdick relates, "static electricity reached out, and he hit the ground." The charge threw the man down, though it left him uninjured. Tex remembers, "It reached out and got him. Now, I'll tell you!"

A Burdick & Burdick crew poses with an unidentified family at a dilapidated wooden windmill tower. Circa 1935.

Without a dependable supply of water, human life in the desert is impossible. Though the sources have varied with peoples and times, during the past century windmill and drilled well have provided reliable supplies of water from underground to thousands of residents in Texas, New Mexico, and Arizona, the region served by Burdick & Burdick.

Speaking of this water, Harry J. Clifford comments, "They almost got to have it or haul it." For families like the one pictured, the windmill often constituted the only source of water reasonably available. As Tex Burdick observes, "Without water, they don't stay there very long. You can't afford to haul water for a family that size. They'd have to have about a hundred gallons at least." Water weighs roughly eight pounds to the gallon, so that would mean hauling eight hundred pounds of water daily.

Rural families during the depths of the Depression often went without even the barest necessities to save enough money to buy a windmill. Even half a century ago windmills were not cheap: a 8-foot Challenge 27 with bronze bearings, probably the right mill for the tower in the picture, cost the customer about $135.00 in 1935.

"It meant life or death to 'em," Burdick says. "They had to have water."

This unidentified worker wearing striped cotton overalls, high lace-up leather shoes, a clean shirt with long sleeves rolled up, and a straw hat examines a wheel section from a sixteen-foot Challenge 27 windmill. Circa 1935.

The members of Burdick & Burdick windmilling crews dressed in much the same manner as laborers anywhere else in the desert Southwest fifty years ago. Their outerwear consisted of overalls or dungarees, which in heavy outdoor work often became soiled. The trousers were usually made of cotton, frequently denim, and shirts also were usually of cotton.

In the El Paso area crew members needed jackets for only a few weeks of the year, but at higher elevations in New Mexico and Arizona, they sometimes needed heavier clothing. Whatever the temperature, most of the men wore hats of one type or another, most frequently made of felt but occasionally of straw. Footwear usually consisted of leather-sole lace-up work shoes often taller than ankle height. Protective steel-toe shoes were unknown. During warm weather, Harry Clifford preferred to go bare chested. "I worked in just a pair of pants and that was all," he said. "I was brown from here up, just black and burnt."

Of all Burdick's men, Bill Berry is remembered today as the most fastidious about his clothing. "He went up on a tower one day and got grease all over his arms . . . and . . . pants," Tex relates, "I mean that old windmill grease." The next thing Burdick knew, Berry was back on the ground and had opened a container of gasoline and was trying to remove the black grease, telling his boss, "If I leave that in there, it'll never come out."

71

Tex Burdick's Challenge 27 demonstrator windmill mounted on a short stub tower in a trailer pulled by an automobile. Circa 1935.

"Get a steel stub tower . . . on which you can place the complete mill . . . and demonstrate it to your prospect customers. It will help you close sales," advised the Challenge Company in its dealer's price list of January 7, 1935. Tex Burdick decided that it was a good idea.

"I built this tower and bolted it to the trailer and put . . . a six-foot mill on it," he says. "I took it out to the dealers and showed how it operated, how it was designed." They could climb right into the trailer, remove the sheet-metal hood, and look down right into the working parts of the mill. Burdick spent two or three months on the road with the trailer promoting sales of the Challenge 27 windmills. Between customers he tied the wheel and vane together so that they would not catch in the wind, and in this way he could travel at the best speed that the mostly unpaved roads would allow.

Taking the Challenge sales literature to heart, Burdick even outfitted a second car with a black leatherette Challenge windmill spare tire cover bearing a Challenge windmill emblem and touted by the company as "one of the most attractive advertisements for the spare tire of your truck or pleasure car."

Twenty-two-foot Challenge 27 windmill on forty-foot steel tower erected by Burdick & Burdick for Bill Brownfield. Lulu, Texas, north of Salt Flat, Texas, near the Texas/New Mexico state line, circa 1935.

Bill Brownfield ran cattle in the scrubby desert just below the western slope of the Guadalupe Mountains north of Salt Flat around the Texas/New Mexico state line. The alkali-whitened land he owned was hardly worth anything. Then Brownfield hired Tex Burdick to put a big twenty-two-foot Challenge 27 over a well that he'd been pumping with an old-fashioned walking-beam-type pump at a place called Lulu.

Initially the windmill operated a 2¾-inch pumping cylinder underground at the water table, but even in the strongest wind the pumping did not lower the water level in the well. "So we put in a bigger cylinder, a 3¾-, and it still didn't go down," Burdick recalls. The performance of the surprisingly strong well beneath the big mill prompted people in the area to begin irrigating with groundwater. By the 1940s Brownfield had sold off much of his rangeland to farmers, who used power pumps to irrigate cotton, vegetables, fruit, and alfalfa, converting the barren flats into a virtual Garden of Eden. By the mid 1960s the farming area around present-day Dell City had a modern school system, seven churches, a medical center, and six cotton gins.

Bill Brownfield, whom Tex remembers as "a crabby old rancher," didn't get to enjoy the proceeds of the sale of his land. "He got lots of money, and he was going to travel around the world," Burdick says. "When he finally sold a big tract of land there and got cash for it, he was going to have just fun, he was going to go out and have a lot of fun," but within eight months he died. "And," Burdick concludes, "the sons . . . and . . . the wife, they fought over it, . . . but that opened up that whole Salt Flat country."

A concrete stock-watering trough and steel water-storage tank with fourteen-foot Challenge 27 windmill and forty-foot steel tower erected by Burdick & Burdick between Salt Flat and the Hueco Tanks. Hudspeth County, Texas, circa 1935.

In the desert Southwest, the most common receptacle for livestock water was the trough, a basin made of masonry, wood, or metal, that caught the water coming either directly from the windmill or from intermediate storage tanks. Those larger storage tanks provided for a constant supply of water even when the wind did not blow. Some of the best troughs were made like the one in the photograph, with the ground surface built up on either side with concrete, stone, or gravel so that it was three to four inches higher than the bottom of the tub.

The unexpected reason for this method of construction, according to Tex Burdick, was the rate of growth of hooves on cattle. "In sandy country . . . the hooves keep growing longer and longer . . . and [for the cattle] it's hard to get around." Their hooves grow faster than they wear away.

The animals found some relief by softening their hooves in water. They would try to step into their watering trough to soak their feet. "They'll soften them up in there, and then they'll walk off some of their hooves," says Burdick. "But if the trough is two or three inches lower on the inside than it is on the outside, she won't put her foot down [inside]."

A typical trough filled automatically from the action of a float valve that allowed water to flow whenever the level in the tubs fell to a predetermined limit. An adjustable outlet kept the storage tank from draining should the float valve in the basin become damaged or stick open. Speaking of the troughs with float valves, former windmiller Harry Clifford remarks, "They saved water, because when they got full, they shut off." With such devices, as the down-to-earth Clifford observes, "nobody'd have to stay there and watch it fill the tank."

Sally and Harold Payne pose with their Bucyrus Armstrong well-drilling rig outside the Burdick & Burdick building at 190 North Cotton Avenue. El Paso, spring 1939.

Harold and Sally Payne were a well-drilling team based in Silver City, New Mexico, where Harold's father had been a driller. They were among the drillers that Tex Burdick recommended to his customers who needed new wells.

If a rancher came into the Burdick & Burdick store at 190 North Cotton Avenue in El Paso and wanted a new well in the area around Silver City, Tex would tell him, "Well, Harold Payne is the man to do it." Similarly, whenever Payne had a customer who needed a windmill and pump installed at a new well, he would refer the customer to Burdick. It was a comfortable relationship between two complementary firms, each taking pride in its work.

Sally and Harold worked together in the 1930s until their business merged into a large-scale well-drilling enterprise that extended operations as far as the El Paso Valley. On one of the jobs Sally was injured, and Tex Burdick remembers hearing that she "screamed every time the wheel turned when they were taking her to the hospital." She recovered from the accident and outlived Harold by many years.

Reflecting on the husband-and-wife team, Tex Burdick observes, "She was a nice lady. I liked her."

A shop-made percussion well-drilling rig mounted on the running gear from an army vehicle sits near a well site. Circa 1935.

Harold Payne's father was a well driller, and both Harold and his brother followed in his footsteps. Harold told Tex Burdick about an incident that occurred early in their careers of punching holes in the ground:

The father discovered a well-drilling rig sitting out in the range country with nobody around. "There were lots of tools there: bits and jars and sockets and things like that," Payne said. The father picked up and carried off all the loose tools from the rig and brought them to the site where Harold was working.

Asked where the driller's tools had come from, the senior Payne replied, "Why, they were over at such and such a ranch." They had need of the tools, but unobserved they returned them to the well rig on the ranch. Harold closed the affair by telling his father, "I don't want anything like that to happen to the Paynes."

A Burdick & Burdick crew prepares to disassemble a twenty-foot open-geared Samson steel windmill at the headquarters of the W. R. Lovelace Ranch. Bill Berry straddles the vane stem, demonstrating his "immunity" to heights. Tecolote, south of Corona, New Mexico, circa 1935.

Over the years, Burdick & Burdick crews performed a number of jobs for W. R. Lovelace on his sheep ranch at Tecolote in Lincoln County, New Mexico. One of the most interesting involved the replacement of a twenty-foot-diameter open-geared Samson windmill on a wooden tower with a new eighteen-foot Challenge 27 on a steel tower at the ranch headquarters.

The job started with the disassembly of the badly worn Samson. Once the old mill was removed, the crew placed the entire wooden tower upright on pipe rollers and moved it a number of feet to one side. Guyed securely, the old wooden tower provided leverage for the cables used to raise the new tower from its horizontal assembly position. Once the new tower was erected and anchored firmly, it served in turn as the vertical support for the cables used to let down the wooden tower for disassembly. The job required several days.

Tex and his men had expected to camp in some comfort at the Lovelace Ranch headquarters. Much to their surprise, however, the wife of the rancher insisted that they stay elsewhere. The windmill men slept almost twenty miles away at the little town of Corona, New Mexico, where they found temporary shelter in an empty boxcar on a railway siding. "Every time a train would come along in the night," Burdick remembers, "we would roll up our beds. We didn't know whether we were going to get moved or not."

The only double-height steel water-storage tank built by Burdick & Burdick. M. A. Fairchild Ranch west of Douglas, Arizona, circa 1942.

Among the services offered by Burdick & Burdick was the construction of custom-built, circular, steel water storage tanks designed to hold reserve supplies of water for times of calm when the windmill did not pump and times when the well was out of service.

Tex Burdick's crews built those tanks from six-by twenty-foot sheets of steel—salvaged from disassembled oil field tanks—that they purchased in El Paso and hauled to the construction sites. In the early years most of the sheets were a quarter or three-eighths of an inch thick and punched with holes every two inches around the edges.

At an installation site some of the crew members would first construct a temporary base of rocks or scrap material to support the circular tank, while other men would start bolting the heavy sheets together end-on-end using tar paper as gasket material. After two or three of the sheets were attached, the workers could start manhandling them into an oval shape, attaching sheets until they could connect the ends. As Harry Clifford recalls, "We just started with brute force. . . . The longer it got, we could bend 'em around till we got the ends together."

Only if a steel storage tank were truly circular and perfectly level, would it achieve its full potential strength. To describe the circle for the tank, the men would sink a wooden post as near the center as possible and drive a metal pin into its top. They would drill a hole in one end of a piece of lumber about as long as the radius of the tank and set the hole over the pin. With a level fastened to the board, the device enabled them to judge the roundness and levelness of the tank. They would adjust the steel sheets with crowbars into proper alignment. "We'd raise it and we'd lower it and we'd get it round," Burdick remembers.

Once the tank was positioned correctly, the men would pour concrete into forms made of scrap lumber to make a base about a foot wide, reinforced with old steel cables, around the bottom outside of the tank. The cables helped prevent cracking of the foundation as the metal in the tank expanded and contracted with changes in temperature. Then the crew would reach over the top of the six-foot-tall steel walls to pour concrete into the bottom of the tank, puddling it on the back sides of the steel sheets and covering the bottom to make it as nearly watertight as possible.

Completing the installation, the men would put together the pipes that delivered water to the tank as well as those that conveyed it to troughs or wherever it might be needed. Then the men would fill the tank with water, just to be certain there were no major leaks, before they headed back to El Paso.

89

Windmiller Carl Boyd (left) and an unidentified Burdick & Burdick employee preparing to install sucker rod in a nine-hundred-foot well beneath a twenty-two-foot Challenge 27 windmill. John Prather Ranch, Alamogordo vicinity, New Mexico, circa 1942.

Among B. H. Burdick's most notable customers was John Prather, owner of a large ranch in the vicinity of Alamogordo, New Mexico. Burdick undertook several jobs for Prather, including the erection of the twenty-two-foot Challenge 27 windmill and steel tower shown here.

Prather had come years before to the Tularosa Valley, and when the United States Army attempted to incorporate his property into the White Sands missile range, he balked at the idea. Prather told Burdick that when military representatives had visited the ranch, he had agreed to give them any specific areas that were needed for missile testing purposes, but had refused to vacate all his lands. "I came here with a set of mules and a covered wagon, and I'm not going to leave here," the old man said.

The army first sent out a sergeant with papers to remove Prather from the land, but the stockman refused to budge. Then they sent out a lieutenant, and finally they sent a major. Every time army representatives came, according to Burdick, "he'd . . . have 'em coffee and dinner and everything like that, but he wasn't going to leave." Prather told Burdick, "If there'd been a [civilian] government man out here, my shirttail wouldn't have touched my rear end before I got off from here, but when you send these army people out here telling me to get out, I'm not going to go."

To say that Prather was stubborn about his rights as he saw them would be an understatement.

When government officials deposited a large sum of money in a special account in the bank at Alamogordo in Prather's name as a means of forcing the purchase of the land, the old man refused even to acknowledge the account. After a few months, the Internal Revenue Service notified Prather that he had not paid any taxes on the funds, to which he responded, "It's not mine. It's yours. I haven't touched it." He didn't want the money; he wanted only to be left alone.

The dispute continued for months. Only after John Prather passed on did the White Sands missile range absorb his ranch. He became a legend in the Tularosa Valley for his persistence in the face of government coercion. Tex Burdick summarizes the standoff well: "He told 'em to go to Hell, and he meant every word of it."

91

A stack of sucker rod and the tools used to pull it from the nine-hundred-foot well at the headquarters of the John Prather Ranch. Alamogordo vicinity, New Mexico, circa 1942.

"Pulling the rods and pipe" was part of the standard maintenance that Burdick & Burdick crews performed on windmills. The procedure consisted of withdrawing the sucker rod and drop pipe from the well, making the repairs needed, and returning them to the well.

Drop pipe is the removable metal pipe in jointed pieces that extends downward into a well from the surface and that usually has a pump mounted at its base below the water table. Sucker rod, composed of segmented poles made of wood or metal, runs within the drop pipe to transmit the up-and-down stroke to the pump. Typically, both sucker rod and drop pipe come in threaded pieces with an average length of about twenty to twenty-one feet. As the sucker rod moves up and down, actuated by the windmill or an engine, valves in the pump open and close automatically to push water up the drop pipe to the surface.

On all but the shallowest wells, windmillers had to employ the leverage of blocks and tackle to raise up rods and pipe and to put them back. On the deepest wells, such as the nine-hundred-foot well on the John Prather Ranch shown here, the combined weight of the rods or pipe was enormous. "You can see all those rods that went into the well, so you can see it wasn't little," remarks Tex Burdick. "And those were the kind of blocks we used. . . . Look at the size of that hook. . . . That's no small job." The work at the Prather Ranch was memorable for the crew members because of the weight of the hundreds of feet of drop pipe that had to be removed and then replaced in the well. "We had to use a truck and a car behind it to pull that pipe [because] it was so heavy," says Burdick.

93

Three pipe elevators and a set of cast-iron bushings in front of the vane sheet from an eight-, nine-, or ten-foot Challenge 27 windmill in the alley outside the Burdick & Burdick building. El Paso, circa 1940.

Pipe elevators are the tools that fasten to blocks and tackle for hoisting drop pipe either into or out of a well. They are designed to clamp around the pipe segments just below the slightly broader couplings. They must clamp tightly enough to hold a pipe securely but loosely enough to allow a worker with a large wrench-like chain tong to unscrew or reattach the individual segments.

In removing drop pipe from a well, the windmiller first fastens the pipe elevator just below the coupling on the upper end of the piece of pipe that protrudes from the well, and then with the blocks and tackle he raises it until the next coupling about twenty-one feet lower comes out of the well. He then attaches a second pipe elevator beneath the newly exposed coupling to keep the string of pipe from falling back into the well. Next with a large chain-tong he unscrews the upper piece of pipe, which is sticking up into the air within the tower. "Usually you use . . . a six-foot chain tong to walk that 'round and 'round the well," advises Harry Clifford. "Three-foot won't budge 'em. . . after they've been set up for a while." After the piece of pipe has been unscrewed, it can be stacked upright inside the tower or pulled out and laid on the ground. The operation then is repeated until all the drop pipe has been removed from the well. To return the drop pipe to the well, the process is reversed.

Pipe elevators were expensive. Those used by Burdick crews in the 1930s, made by the Dempster Mill Manufacturing Company, cost sixteen to eighty dollars a pair, depending on the size. An entire set of elevators for the most commonly used pipe sizes—2, 2½, 3, 3½, and 4 inches—would cost a little under a hundred dollars just over fifty years ago. To save money Tex Burdick had the iron founders at Darbyshire Steel in El Paso fabricate less expensive cast iron bushings that would convert the 4-inch pipe elevator to use in work with 2-, 2½-, 3-, and 3½-inch pipe as well as with the size for which it was designed. "That way," Burdick explained, "a man could have a 4-inch elevator or a 3 or a 2½- or whatever he needed with the same set . . . ."

An unidentified man sitting on the platform of a thirty-nine-foot wide-spread steel tower, made by the Dempster Mill Manufacturing Company, with a new sixteen-foot Challenge 27 windmill. 1940s.

Wide-spread towers were designed with legs spread farther apart at the base than legs of regular towers. Customers might choose such derricks for a number of reasons, but in the desert Southwest it was usually for either convenience in well service or stability in high winds.

A regular-style tower, with its base about a fifth as wide as its height, cramped the working space allowed for servicing the well. Space was especially critical when the windmiller wanted to store drop pipe or sucker rod upright within the tower while working on the pump. "They just wanted room to work underneath there," Tex Burdick observes. Because theoretically the wider placement of anchors gave the steel derricks more stability in extreme winds, some customers would ask Burdick & Burdick for towers of this style for use in areas prone to windstorms. An occasional customer would need a wide-spread tower because only its wider base would fit around the opening of some hand-dug wells.

The selection of a windmill and tower depended on local conditions and the needs of the customer. The most important factor in determining the proper height of a tower was the requirement that the mill project at least fifteen feet above any obstructions. Many customers chose towers at least thirty feet tall in order to provide working space within the tower for pulling sucker rod and drop pipe, which usually came in approximately twenty- to twenty-one-foot lengths. Decisions on the size of the mill were based on the depth of the well and the volume of water required at the surface. The larger mills could pump more water than the smaller ones under the same conditions.

At normal operating speed, the tips of the blades on a windmill wheel move at approximately forty-three miles per hour whatever the diameter of the wheel. To make that speed, Tex Burdick notes, "a little old six-foot mill . . . turns to beat the band." The mill is spinning so fast and producing so many pump strokes that the pump rod is almost jerking up and down. "But you take a twenty-foot mill," Burdick observes, "the outside [of the wheel is] making forty-three [miles per hour, but] the stroke's down to thirteen or fourteen [per minute]." Because of its larger size, the big mill appears to be moving sluggishly although the tips of its blades are moving at the same forty-three miles per hour as those on the little six-footer.

96

Wind-damaged wheel from a fourteen-foot Challenge 27 windmill.   Figure 2 Ranch, western Culberson County, Texas, circa 1935.

Not every Burdick & Burdick installation succeeded.   About 1934 or 1935 Tex Burdick erected a frustrating series of mills on the Figure 2 Ranch at the base of the Sierra Diablo Mountains in western Culberson County, Texas.   His customer there had had problems with wind damage to various brands of windmills and had approached Burdick with his special needs.   Burdick agreed to install a fourteen-foot Challenge 27 on a forty-foot steel tower, and as a matter of pride he guaranteed that the mill and tower would stand.   Perhaps he shouldn't have done so.

The well was at the vortex of an air current caused by a canyon opening out of the mountains. As Burdick recalls, "The wind would come around there in such turbulence that the wheel would spin like a top."   Then suddenly the vane assembly would be blown into the wheel and rebound, and the wheel would spin madly again for a while.

Tex Burdick suffered repeated losses on his guaranteed mill on the Figure 2 Ranch.   First a break occurred in the main shaft, causing the wheel to tumble to the ground and damaging six of its eight sections beyond use.   Repairs were made, but the next storm "blew the tower down and ruined the mill."   Burdick erected a new tower and mill, using what few parts could be salvaged from the wreck and adding double braces to the tower.   The next big windstorm ruined yet another wheel.

Tex had had enough, so, "we took the whole damned works down . . . and . . . we gave the man back the money."   Reflecting on his failure, Burdick salves his self-esteem with the knowledge that his client had already "had five different mills over that well.   None of 'em stood up more than a year."

99

A Burdick & Burdick crew works in a box canyon to replace a pump. Delaware Mountains south of Carlsbad, New Mexico, circa 1940.

"An old cow can only go three days without water," says Tex Burdick, discussing the urgency of repairs to the water systems on ranches. "When a well or a windmill or an engine breaks down, they've got to get the equipment there to repair it within three days, or else they have to haul water." Hauling water soon becomes an onerous burden, or as Tex puts it, "To haul water for a lot of cows that [each] drinks ten to fifteen gallons a day is quite a job." So whenever Burdick & Burdick received an emergency call from a client who had lost an important source of water, the crews of men headed out from El Paso as soon as they could get on the road. Time was at a premium, and road travel could be slow.

The site in the photograph was in a box canyon on the eastern side of the Delaware Mountains south of Carlsbad, New Mexico. The wind turbulence in the Delawares probably had prompted the rancher to choose his J. I. Case diesel engine and Alamo Iron Works back-geared deep well pump jack instead of a windmill, which most likely would have blown down in this location. "They wanted a well pulled," Tex remembers, but there was no tower over the well to support the pulling equipment. Instead, the Burdick & Burdick men brought with them a pair of long pipes, which they chained together at the top to form a temporary vertical support for removing the sucker rod and drop pipe from the well. "We had to pull the pipe and the whole works and put on a new cylinder," Burdick says.

The windmillers remembered the setting as well as the job. Of the bumpy track that they had to take from the Carlsbad highway across the ranch to the remote well, Tex remarks, "It was quite a rough road, let's put it that way."

100

Burdick & Burdick company top man Sid Bowlin (left) with other men undertaking service on a well. Delaware Mountains south of Carlsbad, New Mexico, circa 1940.

Already in the 1930s B. H. Burdick could see that the windmill represented a declining base for his company business. Some competition came from pump jacks operated by internal-combustion engines, but most of it came from electric pumps that operated on power provided since 1935 under the auspices of the Rural Electrification Administration. "The REA went out to all these ranches," Burdick observes. "They might run a line twenty, thirty miles out to a ranch."

The ready source of electric power revolutionized much of rural water supply. Today if they have access to electricity, most ranchers choose to place an electric submersible pump in a strong well and pump water to a storage tank on the highest convenient ground that they have. Tex Burdick explains, "Instead of having

windmills strung out around the country, they will put in . . . [plastic] pipe and run it from this tank to these drinking tubs out maybe a mile or two miles or five miles or even ten miles." Typically the ranchers place each drinking trough within about three miles of another to minimize the distance the cattle must walk to water.

The heyday of the windmill gone, the Burdick & Burdick Company of El Paso, now in the hands of Tex Burdick's son and grandson, remains an active economic force in the region. Though the firm still sells a few windmills each year, it serves many more of its customers with one of the products that reduced the role of the windmill in the desert Southwest: plastic pipe.

102

103

Recommended Readings

The use in America of self-governing wind machines to pump water goes back to 1854 and Daniel Halladay's invention of the first successful such windmill. Wind machines for pumping water began appearing in the desert Southwest in considerable numbers in the 1880s and by the turn of the twentieth century they had become commonplace. In the overall scheme of wind power history, B. H. Burdick was a comparative Johnny-come-lately to the business of providing wind-pumped water to livestock, but even so his firm symbolizes the heyday of windmill use.

A number of different materials can lead readers to a better understanding of the history of wind power in the United States and particularly of the context in which Burdick & Burdick functioned in the desert Southwest. The following sources are recommended for readers who would like to learn more about these subjects.

INTERVIEWS IN THE RESEARCH CENTER, PANHANDLE-PLAINS HISTORICAL MUSEUM, CANYON, TEXAS

Burdick, B. H. "Tex," Sr., to Larry D. Sall. El Paso, Texas, December 12, 1975. Tape recording and typescript.

——, to T. Lindsay Baker. El Paso, Texas, March 6, 1981. Tape recording and typescript.

——, to T. Lindsay Baker. Rio Vista vicinity, Texas, March 14, 1991. Tape recording and typescript.

Clifford, Harry J., to T. Lindsay Baker. Albuquerque, New Mexico, June 16, 1991. Tape recording and typescript.

GENERAL AMERICAN WINDMILL HISTORY

Baker, T. Lindsay. "Challenge Line of Windmills." *Windmillers' Gazette* (Canyon and Rio Vista, Tex.) 2, no. 3 (Summer 1983): 3-8.

——. *A Field Guide to American Windmills.* Norman: University of Oklahoma Press, 1985.

——. "New Deal Specials: Bargain Windmills of the Depression Era." *Windmillers' Gazette* 4, no. 2 (Spring 1985):3-5.

——. "Turbine-type Windmills of the Great Plains and Midwest." *Agricultural History* 53, no. 1 (January 1980): 38-51.

Davis, Mary Margaret. "Windmill Company in Updraft: Business Goes in Circles." *El Paso Times* (El Paso, Tex.), August 18, 1981, Sec. C, p. 1.

Dick, Everett. "Water: A Frontier Problem." *Nebraska History* 49, no. 3 (Autumn 1968): 215-45.

Eide, A. Clyde. "Free as the Wind." *Nebraska History* 51, no. 1 (Spring 1970): 25-48.

McBrinn, Bob. "Events Took Sales Out of Windmills." *El Paso Herald-Post* (El Paso, Tex.), February 5, 1968, Sec. B, p. 1.

Torrey, Volta. *Wind-catchers: American Windmills of Yesterday and Tomorrow.* Brattleboro, Vt.: The Stephen Greene Press, 1976.

"Windmills and Wind Engines." *Farm Implement News* (Chicago), 8, no. 1 (January 1887): 13-15; no. 2 (February 1887): 12-15; no. 3 (March 1887): 12-15.

"Wind-power—Its Utilization and Various Applications." *Farm Implement News* 13, no. 10 (March 10, 1892): 13-18.

Wolff, Alfred R. *The Windmill as a Prime Mover.* New York: John Wiley & Sons, 1885.

Hays, Dick, and Bill Allen. *Windmills and Pumps of the Southwest.* Austin, Tex.: Eakin Press, 1983.

Hirschberg, Gary. *The New Alchemy Water Pumping Windmill Book.* Andover, Mass.: Brick House Publishing Company, 1982.

McKenzie, Dan W. *Range Water Pumping Systems: State-of-the-Art-Review.* Forest Service Equipment Development Center Project Report 8522 1201. Washington, D. C.: United States Department of Agriculture, 1985.

"Moves with a Breath of Air." *Cattleman* (Fort Worth, Tex.) 29, no. 3 (August 1942): 5-6.

Nelson, Bascom. "The Windmill and the Cattle Industry." *Cattleman* 54, no. 1 (June 1967): 41, 73-74.

"Thousands Invested in Windmills." *Sheep and Goat Raisers' Magazine* (San Angelo, Tex.) 10, no. 4 (November 1929): 110, 112.

RANCH WATER SUPPLY

Baker, T. Lindsay. "Windmills of the Panhandle-Plains." *Panhandle-Plains Historical Review* (Canyon, Tex.) 53 (1980): 71-110.

Barnes, Will C. *Stock-watering Places on Western Grazing Lands.* U. S., Department of Agriculture, Farmers' Bulletin no. 592. Washington, D. C.: Government Printing Office, 1914.

Hamilton, C. L., and Hans G. Jepson. *Stock-watering Developments: Wells, Springs, and Ponds.* U. S., Department of Agriculture, Farmers' Bulletin no. 1859. Washington, D. C.: Government Printing Office, 1940.

HISTORIC WINDMILLERS

Coates, Gary. "Windmill Builder Remembers Past." *San Angelo Standard-Times* (San Angelo, Tex.), March 25, 1973, Sec. B, p. 1.

Hay, Gerald. "Brothers Help to Keep Windmills Churning." *Hutchinson News* (Hutchinson, Kans.), December 5, 1976, p. 35.

Hendrix, John M. "Windmill Monkeys." *Cattleman* 25, no. 8 (January 1939): 51-52.

Kirk, Margaret. "The Windmill Man." *Western Livestock and the Westerner* (Denver) 37, no. 2 (September 1951): 16, 98-99.

Wildenthal, Bryan. "Windmill Schmidt." *Sheep and Goat Raiser* (San Angelo, Tex.) 31, no. 4 (January 1951): 12-14, 48, 50-51.

CONTEMPORARY TRADE LITERATURE FOR
CHALLENGE WINDMILLS AND AUXILIARY
WATER SUPPLY EQUIPMENT

Challenge Company, Batavia, Ill. *Catalog No. 41[,] a Complete Line of Farm and Home Plumbing and Water Supply Equipment.* Batavia, Ill.: Challenge Company, [ca. 1940]. 104 pp. Available in Research Center, Panhandle-Plains Historical Museum.

———. *Challenge Wind Mills Take Dollars from the Air and Put Them in Your Pocket.* Publication 948. Batavia, Ill.: Challenge Company, [ca. 1930]. 4 pp. Loc. Cit.

———. *Dealers' Price List[,] Challenge Windmills and Steel Towers No. 35 [,] January 7, 1935.* Publication 1080. Batavia, Ill.: Challenge Company, 1935. 8 pp. Loc. Cit.

———. *General Catalog No. 92[,] Challenge Company[,] Manufacturers of Windmills, Towers, Steel and Wood Tanks, Gasoline, Gas and Kerosene Engines, Feed Grinders, Corn Shellers, Wood Saws, Pumps, Cylinders, and General Water Supply Goods, Steel Sub-structures and Water Works Supplies.* Aurora, Ill.: Strathmore Co., 1929. 224 pp. Loc. Cit.

Challenge Company, Kansas City, Mo. *Net Prices on "Challenge 27" Self Oiling Windmill [,] February 26th[,] 1931.* Publication 1049. Kansas City, Mo.: Challenge Company, 1931. 1 lf. Loc. Cit.

Challenge Company, Omaha, Neb. *Wholesale Price List Applying to General Catalog No. 92 [,] June 1, 1929 [,] No. 29 A.* Publication 1017. Omaha, Neb.: Challenge Company, 1929. 40 pp. Loc. Cit.

INSTALLATION AND MAINTENANCE OF
WINDMILLS

Baker, T. Lindsay. "Anchors for Windmill Towers." *Windmillers' Gazette* 3, no. 4 (Autumn 1984): 9.

———. "Annual Lubrication for Self-oiling Windmills." *Windmillers' Gazette* 4, no. 1 (Winter 1985): 7-8.

———. "Building Steel Towers Up from the Ground." *Windmillers' Gazette* 5, no. 3 (Summer 1986): 6-8.

———. "How to Build a Wooden Windmill Tower." *Windmillers' Gazette* 6, no. 4 (Autumn 1987): 6-7.

———. "How to Rawhide a Wooden Windmill Wheel." *Windmillers' Gazette* 3, no. 2 (Spring 1984): 9-10.

———. "How to Replace Babbitt Windmill Bearings." *Windmillers' Gazette* 2, no. 2 (Spring 1983): 8-10.

———. "How to Use a Gin Pole in Installing Windmills." *Windmillers' Gazette* 8, no. 1 (Winter 1989): 4-6.

———. "Raising Fully Assembled Windmill Towers." *Windmillers' Gazette* 5, no. 4 (Autumn 1986): 6-9.

———. "Raising Wooden Windmill Towers." *Windmillers' Gazette* 7, no. 1 (Winter 1988): 5-7.

Dickerson, I. W. "Care of Windmills." *Wallace's Farmer* 43, no. 4 (January 25, 1918): 126, 128.